Mathematics without Numbers

Mathematics without Numbers

Towards a Modal-Structural Interpretation

GEOFFREY HELLMAN

CLARENDON PRESS · OXFORD

1989

Oxford University Press, Walton Street, Oxford OX2 6DP

Oxford New York Toronto
Delhi Bombay Calcutta Madras Karachi
Petaling Jaya Singapore Hong Kong Tokyo
Nairobi Dar es Salaam Cape Town
Melbourne Auckland
and associated companies in
Berlin Ibadan

Oxford is a trade mark of Oxford University Press

Published in the United States
by Oxford University Press, New York

British Library Cataloguing in Publication Data
Hellman, Geoffrey
Mathematics without numbers: towards a
modal-structural interpretation
1. Mathematics
I. Title
510
ISBN 0-19-824934-9

Library of Congress Cataloging in Publication Data
Hellman, Geoffrey.
Mathematics without numbers : towards a modal-structural
interpretation / Geoffrey Hellman.
Bibliography: p.   Includes index.
1. Mathematics—1961–   I. Title.
QA37.2.H42   1989   510—dc19   89-30517
ISBN 0-19-824934-9

Typeset by Eta Services (Typesetters) Ltd., Beccles, Suffolk
Printed and bound in
Great Britain by Biddles Ltd,
Guildford and King's Lynn

TO THE MEMORY
OF MY PARENTS

Preface

> ... mathematics may be defined as the subject in which we never know what we are talking about, nor whether what we are saying is true.
>
> *Bertrand Russell*

THE idea that pure mathematics is concerned principally with the investigation of structures of various types in complete abstraction from the nature of individual objects making up those structures is not a novel one, and can be traced at least as far back as Dedekind's classic essay, 'Was sind und was sollen die Zahlen' (originally published in 1888). It represents a striking contrast with the Fregean preoccupation with identifying "the objects" of particular branches of mathematics, and it seems to have lain behind Hilbert's refusal to accept Frege's point of view on such fundamental matters as the nature of mathematical definitions and axioms, mathematical existence, and truth.

With the rise of the comprehensive "logicist" systems of type theory and axiomatic set theory, however, the structuralist idea was either neglected in favour of some arbitrarily chosen relative interpretation of ordinary mathematics (number theory, analysis, etc.) within the comprehensive system, or else it was given metalinguistic lip-service through the apparatus of Tarskian model theory, carried out within set theory itself.

Despite the attractive unifying power of modern set theory, embedding the structuralist intuition within it has its disadvantages. Must we accept anything so powerful as, say, the Zermelo–Fraenkel axioms—categorically asserted as truths about Platonic objects—in order to carry out a structuralist interpretation of number theory or classical analysis? And what of set theory itself; can it not also be understood along structuralist lines, and would it not constitute a philosophical advance so to understand it? "It", after all, is actually a multitude of apparently conflicting systems. On the standard Platonist picture, at most one of them can correctly describe "the real world of sets". On a structuralist interpretation, there is at least the prospect of a healthy pluralism—that many different systems can be sustained as theories investigating different structural possibilities.

A further main source of inspiration for the present study is Hilary Putnam's 'Mathematics without Foundations' [1967], which suggested that modal logic could be used as a framework for eliminating apparent reference to mathematical objects entirely, including the objects of abstract set theory itself. Conundrums associated with a special realm of mathematical objects, emphasized by a number of contemporary philosophers such as Nelson Goodman and Paul Benacerraf in terms strikingly reminiscent of Dedekind—how to reconcile talk of such objects with the multiplicity of "ways of taking them", how we ever manage to refer to such objects, and the like— such questions would be seen *not even to arise* on the modal logical eliminative interpretation. But the details and implications of such an interpretation have remained to be worked out.

The present work is an exploratory effort at synthesizing these two strands of thought, "structuralism" and "mathematics as modal logic". The aim has been to provide for structuralism frameworks that are flexible, suitably powerful, and as precisely delineated as those that have grown out of the logicist tradition and which have dominated non-constructive foundational thought.

When I began working on this project several years ago, Putnam was my principal source for the idea that set theory itself ought to be understood along structuralist lines and that such an approach contains the germ of a resolution of the problem of proper classes. It was something of a revelation when I stumbled upon a classic opus of Zermelo, *Über Grenzzahlen und Mengenbereiche* [1930], containing strikingly similar proposals. Although this work is known to mathematicians for its leading mathematical content (including the "discovery" of inaccessible cardinals), its intriguing philosophical perspective has yet to receive the attention it deserves. (Unfortunately, an English translation has yet to be published. Doctoral language requirements in this area still have their point!)

Concerning personal acknowledgements, I owe deep and lasting gratitude to Nelson Goodman for years of intellectual guidance and inspiration, although I am resigned to his finding all too little evidence of this in the present work. More proximately, I am indebted to many people for encouragement, criticism, suggestions, and help. Among them are students in graduate seminars in philosophy of mathematics that I have offered at Indiana University and the

University of Minnesota, and members of many audiences at conferences and colloquia who have endured earlier versions of various parts of this work. I am especially grateful to the following individuals: John Burgess, Jeremy Butterfield, Nino Cocchiarella, J. Alberto Coffa, William Craig, J. Michael Dunn, Hartry Field, Richard Grandy, Michael Hallett, W. D. Hart, Daniel Isaacson, Ronald Jensen, Akihiro Kanamori, Saul Kripke, Penelope Maddy, Charles Parsons, Hilary Putnam, Jack Silver, Howard Stein, W. W. Tait, and, in all probability, a number of others for whose understanding I must now plead.

Much of the work for this monograph was carried out with the support of the National Science Foundation, Grants No. SES-8420463 and No. SES-8605286, for which I am very grateful.

G.H.

Minneapolis, Minnesota
8 July 1988

Contents

Introduction

If the queen of the sciences occupies her throne because of the wealth and beauty of her results, where if not among the lowly peasantry are we to locate the philosophy of mathematics? But social categories do not fix the physical surroundings, and it would be misleading to picture this humble being as firmly rooted to the land. Better, perhaps, the image of a forlorn fishing-vessel, doomed to chart its daily course between Scylla and Charybdis—its principal "isms"—between its "platonism", so satisfying in its respect for mathematical truth but so problematic in its treatment of our mathematical knowledge, and its "constructivism", representing a mirror-image situation, with "satisfying" and "problematic" interchanged.

Is there any hope (through meditation, presumably) of our protagonist's breaking out of this wheel of existence? Is it possible to develop an interesting alternative view in the philosophy and foundations of mathematics which somehow retains the main advantages of the existing, antagonistic positions, while at the same time managing to improve significantly on those positions by avoiding their principal weaknesses? Our overarching purpose in this work is to explore this: to lay the groundwork for such an alternative by developing the formal and philosophical core of a view that synthesizes certain "structuralist" ideas (traceable to Dedekind) and some intriguing suggestions (of Putnam) concerning "mathematical possibility and necessity", a core that can be further developed in various ways, but which seems promising philosophically and, moreover, is very interesting in its own right.

But before delving into these ideas and their development, it would be desirable to agree, if possible, on the conditions that prevail in Nirvana, on what we might ideally hope for from a philosophy of mathematics. Now, I am acutely aware that one person's paradise is another's inferno, and that it is scarcely possible to lay down desiderata for a philosophy of mathematics without appearing to be begging crucial questions. I will assume this risk, however, in the belief that a brief list of desiderata will help motivate our inquiry and that it will at least enable the reader to take into account what may be my own prejudices.

The first desideratum rests on the view that the queen merits the full respect of her handmaid: it is desirable to uphold the objectivity of as much mathematics as possible. By "objectivity" here is meant determinateness as to truth value—true or false—, and truth for the error-free portions; moreover, such truth is to be understood classically as implying independence of the particular mathematical investigators. Even if no mathematicians had ever existed, still it would have been the case that every infinite bounded set of reals has (would have had) an accumulation point, etc.

This first desideratum might be summed up by saying that a philosophical interpretation of mathematics should be *"realist"*, as opposed to *"instrumentalist"*. This is one important dimension along which various interpretations of mathematics can be arranged. Traditional platonist views, of course, count as *realist*: mathematical discourse is understood as consisting of statements or propositions that have determinate truth value, independent of our minds. Varieties of formalism, finitism, and constructivism that regard all or parts of classical mathematics as lacking in truth value count as (more or less) *instrumentalist*. It is crucial, however, from our point of view, that this dimension should not automatically be conflated with the contrast between *"platonism"* and *"nominalism"*. This second dimension concerns whether such truth as is recognized holds in virtue of a realm of abstract objects (numbers, sets, functions, etc.) which forms the subject-matter of mathematical discourse. Of course, if one simply reads ordinary mathematical discourse literally, i.e. takes it "at face value", one arrives at a platonist interpretation, one that is also realist. (Various such interpretations are possible, depending on just what counts as "ordinary mathematical discourse". It might be taken as already "reconstructed" as set-theoretic discourse, for example, or it might not be. At this stage, such distinctions do not matter, but at various junctures they must be brought into play.) *But it must not be assumed that, conversely, all realist interpretations must be platonist.* On the contrary, we make no such identification, for we seek an alternative, non-literal interpretation of mathematical discourse which can be understood as realistic but in which ordinary quantification over abstract objects is eliminated entirely. To identify realism with platonism at the outset is to assume in advance that no such interpretation is possible.

A second desideratum concerns epistemology: a philosophical interpretation of mathematics ought to admit of an extension that reasonably accounts for how we come to know or justify that mathematics which we can reasonably be claimed to know or be capable of knowing. Here I assume that appeals to an irreducible "intuition"— to explain, for example, how we have access to a realm of abstract objects—are unsatisfactory (much as appeals to "esp" in other contexts are unsatisfactory). Of course, it must not be assumed that all true mathematical assertions are knowable, even in principle. Nor should it be assumed that "knowledge" or "justification" is an all-or-nothing affair. Rather, we incline to the view that mathematical knowledge, like ordinary and scientific knowledge generally, comes in degrees and may admit of "justification" of diverse kinds. We do not wish to prejudge these issues. But what this desideratum calls for is a reasonable integration of mathematical knowledge with the rest of human knowledge, something that traditional platonist interpretations have difficulty in providing. (Here one should distinguish traditional platonist interpretations, which take mathematical knowledge as absolute and a priori, from the more recent holistic, quasi-empirical platonism of Quine,[1] according to which pure mathematics receives its justification through its empirical, scientific applications. On our own view, this approach has provided valuable insights, and, in fact, we will suggest adapting aspects of it to the interpretation to be developed below.)

Perhaps the central puzzle raised by traditional "objects-platonism" (as I will sometimes explicitly call it, to distinguish a kind of view that goes beyond *realism* in its explicit ontological commitments) is the difficulty in seeing how it is that the posited abstract objects "play any role"—"make any difference"—in our knowledge and in our language. Prima facie, at least, it is difficult to understand, for example, how it is that we justify construing mathematical reference as reference to particular abstracta, as opposed to others forming a structurally isomorphic system—or how, for that matter, such

---

[1] This standpoint on mathematics results from joining two recurrent themes in Quine's corpus, his holistic view of science, on the one hand, and his logistic approach to ontology, on the other. See e.g. 'Two Dogmas of Empiricism' (1951) in Quine [1961], pp. 20–46, esp. pp. 45–6; 'On Carnap's Views on Ontology' (1951) in Quine [1976], pp. 203–11, esp. p. 211; 'A Logistical Approach to the Ontological Problem' (1939) in Quine [1976], pp. 197–202; and 'On Multiplying Entities' (1966) in Quine [1976], pp. 259–64.

reference ever gets established in the first place.[2] Now indeed, there is not even general agreement as to whether such puzzles are genuine theoretical problems or whether, in fact, they are pseudo-problems.[3] All the more reason, then, why it would be desirable to be able to bypass such questions entirely. From a mathematical point of view, questions concerning the absolute identity of "objects of reference" seem entirely irrelevant to mathematical inquiry. Questions, such as, "How do you know to which ω-sequence you are referring when you speak of 'the natural numbers'?" seem utterly alien to the discipline. (In fact, many working in the foundations of mathematics shun such questions entirely. Perhaps it was in reaction to them that Kreisel was prompted to remark that it is the objectivity of mathematics— not mathematical objects—that counts.[4] The wrong emphasis here leaves philosophy of mathematics open to the Allenesque caricature which has it asking, "Do numbers exist? And why? And must they be so silent?")[5] Surely, it would be desirable to understand mathematical discourse in such a way that such questions are completely blocked from the start.

Related to all this is a third desideratum: the prima-facie a priori status of mathematics must be accounted for, if not upheld. In particular, the primacy of proof as the avenue to mathematical knowledge *par excellence* must receive a natural explanation, even if other avenues are left open. One would hope further that the line between the conventional and the non-conventional ('analytic' vs. 'non-analytic') should be clarified. And, it would be pleasant if this could be done with the following happy result: mathematical principles, often taken as axioms, such as mathematical induction or comprehension axioms (restricted in type), of which we should really wish to claim full knowledge, can in some sense be understood as "analytic" or "almost analytic"; yet not all mathematics should turn out empty,

---

[2] For discussion of some of the puzzles associated with objects-platonism, see e.g. Benacerraf, 'What Numbers could not be' (1965) and 'Mathematical Truth' (1973) in Benacerraf and Putnam [1983], pp. 272–94 and pp. 403–20, respectively; and Dummett, 'Platonism' (1967) in Dummett [1978], pp. 202–14. For an approach to such problems on behalf of platonism, see e.g. Maddy [1980].

[3] For a critique of these "problems with Platonism" as not genuine, see Tait [1986*a*].

[4] A remark of Kreisel's along these lines is cited by Putnam in his 'What is Mathematical Truth' in Putnam [1975], pp. 60–78 at p. 70.

[5] "Is there anything out there? And why? And must they be so noisy?" Allen [1971], p. 28.

or true merely by linguistic convention. (For our queen must not turn out to be a mere figurehead!)

Let us pursue this point just a bit further here. We have already spoken of one type of "bad questions"—questions concerning knowledge of unique absolute reference to mathematical objects—that it would be desirable to block. A clarification of the role of convention can help dispose of a second type: questions of the sort just alluded to in connection with certain axioms, such as "How do you know that the reals are densely ordered?", or even, "How do you know that the sets (*sic!*) are well-founded?" There seems to be something badly misguided about such questions, and yet, on standard objects-platonist accounts, it is difficult to dismiss them. (The platonist who reconstructs analysis within set theory can reply to such questions about "the reals" by saying, "I've derived the property in question from the axioms of set theory", but it is much less clear what can be said when the question, such as our second example, concerns an axiom of set theory itself (on standard formalizations).)[6] On a structuralist view, however, there is at least the prospect of a satisfactory answer to many such questions, for, on such a view, the mathematician claims knowledge of structural relationships on the basis of proofs from assumptions that are frequently taken as *stipulative of the sort of structure(s) one means to be investigating.* Thus, if asked to justify an assumption—such as well-foundedness of sets—which can be assigned this stipulative role, there is a ready answer: "These are the sorts of structures I mean to be investigating. One is free, of course, to investigate others." But, how far can such an approach be taken? To which "How do you know" questions is this "conventionalist" response appropriate? Our examples of density of the reals and well-foundedness of sets seem to be clear cases in which it is appropriate, but what about the least-upper-bound principle? Here the answer is less obvious, for we are perhaps prepared to recognize some value in the set-theoretic reduction of analysis (in which the l.u.b. principle is derived from "more basic" set-theoretic principles). Thus, there appears here to be a tension between "structuralism" and "logicism" which will eventually have to be resolved.

[6] On a very elegant alternative formalization based on cumulative levels, due to Scott [1974], well-foundedness can be derived as a theorem. This does not affect the point at issue here, except in so far as it reminds us that very different axiom systems can turn out to be equivalent, so that, on a structuralist view, there are, in general, many different ways of stipulating what sort of structures one is interested in investigating. Cf. below, Ch. 2, § 2.

The fourth and final desideratum I shall mention concerns mathematics in its applications: a reasonable account should be forthcoming of how mathematics does in fact apply to the material world. Here, platonism fares quite well, since it recognizes applied mathematical objects—sets of material objects, functions defined on them with numbers as values, etc.—in terms of which whole mathematical-physical theories can be (and standardly are) understood. It is, rather, constructivism that encounters special problems here. To cite one example: it is well known that the intuitionistic continuum is not totally ordered.[7] But, if we regard a line in physical space, or a stretch of physical time, as totally ordered (as we normally do), how are we to carry out physical geometry? (This is symptomatic of a general difficulty: the quantifiers of mathematical physics do not readily lend themselves to constructive interpretations, even if those of certain portions of pure mathematics do.)[8] This not to say that such problems are insurmountable (from within a constructive framework). But it does point up the importance of addressing the question of application explicitly, and cautions us against the assumption that, if only we can get a scheme or system of interpretation to work for pure mathematics, applied mathematics will "take care of itself".

These, then, are our guiding desiderata. Now we must take up the challenge of trying to forge a view that can meet them.

As some of the discussion thus far already suggests, some kind of "structuralism" seems promising. On the sort of view we should like to articulate, *mathematics is the free exploration of structural possibilities, pursued by (more or less) rigorous deductive means.*[9] This is an intriguing motto, perhaps, but how is it to be developed? What are its basic assumptions, and to what extent can it provide an alternative to traditional objects-platonism? What light can it shed on problems of justification, on the question of analyticity, and on the matter of distinguishing real problems from pseudo-problems? Can such a view be made to work for applied mathematics? And what are the main problems it must confront?

A general structuralist outlook has been widely shared,[10] and can be found in embryonic form, at least, in Dedekind's classic essay,

---

[7] See e.g. Dummett [1977], pp. 47 ff.

[8] This point is made by Putnam [1975], op. cit. n. 4, p. 75.

[9] A point of view along these lines can be discerned in Stein [1988].

[10] For discussion and bibliography, see e.g. Shapiro [1983a].

'Was sind und was sollen die Zahlen'.[11] As I see it, however, there have been two main obstacles to the development of "structuralism". The first is the lack of an explicit, sufficiently precise interpretation of specific mathematical theories. This involves at least the following three things: (i) presenting a translation of the mathematical discourse in question (as commonly represented in quantificational languages) into a specified "structuralist language"; (ii) explicit articulation of the structuralist theory, i.e. the basic assumptions of structuralist mathematics (at least, in so far as it purports to serve as a framework for the given mathematical theories in question—we need not seek a single "universal structuralist theory"); and (iii) a justification of the representation, in so far as that is possible. Until all this is attempted, we remain pretty much in the dark as to the potential of structuralist interpretations.

The second problem, as I see it, is that, as usually presented, it is difficult to see in structuralism any genuine alternative to objects-platonism. This is most obvious when the structures are taken as set-theoretic models, i.e. when the structuralist theory is just set theory (perhaps with urelements), or as members of a category (the theory being category theory, taken literally as quantifying over abstract objects called categories). But this worry also pertains to other attempts (e.g. mathematics as a science of "patterns", where these are taken as platonic entities in their own right).[12]

In what follows, we seek to overcome these obstacles by combining insights of structuralism with a related, more recent strand of thought in philosophy of mathematics, roughly summed up as the view of mathematics "as modal logic". The *locus classicus* of this approach is Putnam's 'Mathematics without Foundations',[13] in which it was suggested that many of the problems plaguing objects-

---

[11] Dedekind [1901]. In precisely what sense Dedekind was a "structuralist" is problematic, as is brought out in the combination of Parsons [1983], Essay 1, Tait [1986b], Parsons [forthcoming], and Stein [1988]. We will return to this matter briefly in Ch. 1, below, where different structuralist theses are distinguished for purposes of such an assessment.

[12] See e.g. Resnik [1981]. In addition to whatever problems there may be in countenancing patterns as universals, there is a special difficulty that arises in the attempt to treat individual constants in standard mathematical languages as referring to "positions in a structure". Simply put, how is one to identify a "position" in a unique way? In the case of natural numbers, this is no problem since, between any two structures of the appropriate type, there is a unique isomorphism. But in the cases of, say, the rationals or the reals, the order-preserving correspondences are not unique—there are infinitely many of them. What then is "the position" picked out by '.33333 ...'?

[13] Putnam (1967), in Putnam [1975], pp. 43–59.

platonism (and, in particular, the identification of mathematics with set theory) could be overcome by reinterpreting mathematics, as standardly presented, in a modal language, in which a notion of mathematical or logical possibility is taken as primitive. Since the work of Kripke,[14] we have become familiar with the procedure of providing a set-theoretic semantics for the operators of modal logic. What Putnam proposed was the opposite procedure of working with those operators as primitive, and using them to reconstruct platonist discourse in such a way that literal quantification over abstract objects could be made to disappear entirely. Some suggestive examples of how this might be done were provided; especially intriguing was the idea of reconstructing Zermelo set theory in terms of a rather complicated translation pattern in which one spoke only of what would be the case in possible extensions of "concrete structures" of the appropriate type, that is, models for Zermelo set theory which were to be characterized somehow by employing "nominalistically acceptable" primitives (e.g. one could speak of marks and arrows, but one could not employ ∈, set membership). Moreover, such a translation pattern was thought to provide an alternative picture that was nevertheless "fully equivalent mathematically" to standardly presented mathematics. However, intriguing as all of this was, and while it did inspire some interesting formal developments in the area of modal set theories,[15] little has been done to develop the original ideas of 'Mathematics without Foundation', that is, to develop explicit translation patterns of mathematical theories into suitable modal theories—capable of standing independently of set theory—and then to justify these as "equivalent for mathematical purposes". Like structuralism, the idea of "mathematics as modal logic" has remained at the level of some seemingly promising suggestions, but it has not been developed even to the point at which a serious philosophical assessment would become possible. One of our goals in what follows has been to remedy this situation.

Our strategy will be to deal in detail with certain specific mathematical theories: to provide for them an explicit "modal-structural interpretation" ("msi") in which literal reference to mathematical objects is entirely eliminated; to articulate the underlying assumptions behind the interpretation; to investigate the question of "equivalence (of the reconstruction) for mathematical purposes";

---

[14] Kripke [1963].
[15] See e.g. Parsons [1983], Essay 11, and Fine [1981].

and to explore some of the main formal and philosophical implications of the reconstruction. We treat essentially three theories: Peano arithmetic and real analysis (in Chapter 1), and Zermelo–Fraenkel set theory (in Chapter 2). These have been chosen both because of their familiarity and because of their central importance, especially from a foundational perspective. (In each case, as is well known, a straightforward formalist or deductivist treatment is ruled out by the Gödel incompleteness theorems: no consistent formal system can generate all sentences standardly interpreted as truths "about the intended type of structure(s)".) However, the approach could well be applied to other mathematical theories, and this should be borne in mind.

In the case of set theory, a modal-structural interpretation along lines suggested by Putnam [1967] has natural applications to "higher axioms" of set theory, especially so-called "large cardinal axioms". This is intimately bound up with problems concerning "proper classes" and provides, from our point of view, one of the most interesting and potentially fruitful applications of the present approach.

In this connection, it may be of some historical interest to note that there are striking similarities between the Putnam-inspired approach to set theory to be developed here and some of the views of Zermelo's classic [1930] paper. The formal content of this late work of Zermelo's (as yet, unfortunately, not published in English translation) is well known among set theorists. Yet its philosophical perspective has yet, I believe, to receive the attention it deserves.

In developing the interpretations, we have made free use of already well-developed axiom systems. Yet we sometimes speak as though we were "representing mathematical discourse", as if the mathematician normally spoke in formal languages. Of course, we make no such assumption. Rather, we assume that standard axiom systems are already a more or less successful codification of significant portions of actual mathematical practice, successful enough to be taken seriously as a starting-point in foundational work. (Here we share the outlook of many mathematical logicians.)[16] If one further desires to relate our reconstruction back to less formalized practice (the "raw data", as it were), one will have further work to do. Although we will sketch a kind of "recovery" of "ordinary practice",

---

[16] See e.g. Feferman [1978], with which we are in sympathy.

this is only part of the story. In particular, one would seek to supplement our account by showing that sense can be made of actual practice which antedates the explicit formulation of axiom systems. That is a task which we have not undertaken here. In contemplating this, however, due consideration must be given to the slack that always exists between informal practice and formal reconstruction, whatever the theoretical inquiry may be. The formal reconstruction is designed to serve certain theoretical purposes, especially to exhibit ways in which various assumptions and methods *can* be justified. It should not be read as asserting that "this is the way mathematics has been (or should have been) carried out".

A final cautionary preliminary: our standpoint throughout will be "non-constructive" in the sense that we freely employ the notion of "arbitrary subset of X" where X may be infinite (or we employ some essentially equivalent notion in point of mathematical richness), and, more fundamentally, we allow—and even seek a kind of justification for—talk of "completed infinite totalities". We see no convincing reason not to take such notions as "meaningful", although we do not enter into anything like a full-scale defence of this classical stance here. As will emerge, some "constructivist" criticisms of this stance may have to be revised in light of modal-structural alternatives to objects-platonism, but we do not pretend to have resolved this central controversy.

It must be said, however, that, while we regard constructive mathematics as part of classical mathematics, and not as a conflicting replacement,[17] and while we have not entered into constructive systems in this monograph, we do see great value in their development from a philosophical perspective. For they can be read as spelling out in detail just how much can be achieved by more or less limited (and correspondingly secure) means. And that in itself is surely a large part, if not the whole, of mathematical epistemology. Although consideration of constructive systems transcends the scope of this study, it may be that certain of the ideas developed here can be adapted to such systems, and that this could turn out to be valuable, especially in connection with the problem of finding support for those modal-structural postulates that prove in the end to be unavoidable.

[17] Here we share the perspective of Tait [1983].

1

The Natural Numbers and Analysis

§0. Introduction

The natural numbers form a natural starting-point for this study—and in more than one way. Not only are they the most fundamental and familiar objects of mathematical thought and practice; they also lend themselves most readily to a structuralist interpretation according to which they—as particular abstract objects—are dispensed with entirely. It is a widely, if not universally, accepted view that, in the theory of arithmetic, what matters is structural relations among the items of an arbitrary progression, not the individual identity of those items. As one commonly says: "Any ω-sequence will do."

It is one of the great ironies of our subject that this view—one that is implicitly recognized in modern set-theoretic treatments of ordinals, where quite arbitrary stipulations are adopted and recognized as such—was explicitly rejected by Russell,[1] who championed a particular way, due principally to Frege,[2] of reducing number theory to a theory of classes. That approach certainly had its attractions (and still does).[3] Russell was particularly enamoured of the ease with which it accommodated ordinary counting: the relation between the enumerated class and the number counted was simply membership. (The oriental kings were three on account of the class of those kings *belonging* to 3.) And, Russell argued, if the natural numbers were identified arbitrarily with "any old ω-sequence", you couldn't account for counting.[4] One might read Russell charitably here as saying that you couldn't *as elegantly* account for counting. For surely you *can*—and you *do* in standard set theoretic treatments—by asserting the existence of a bijection between the class enumerated and (say) the predecessors of the number reached (0 counting as the first number). This is slightly more long-winded

---

[1] In Bertrand Russell [1919], pp. 9–10.

[2] In Frege (1884) in [1978].

[3] For a recent formalization of the mathematical core of Frege's *Die Grundlagen der Arithmetik* in a second-order system provably consistent relative to classical analysis, see Boolos [1987].

[4] Russell [1919], n. 1.

than the Frege–Russell account, to be sure. But, as it happens, there are overwhelming advantages to the modern set-theoretic treatments, as compared with Russell's, when the need to impose type restrictions on classes (or propositional functions) to avoid the paradoxes is taken into account. (For then, Russell's numbers— classes of n-fold classes of a given type—are reduplicated at all levels of the type hierarchy. And on a modern set-theoretic reconstruction, Russell's numbers turn out to be proper classes—"too big" even to be found in Zermelo–Fraenkel set theory, and "too big" to be collected in Von Neumann–Bernays–Gödel set theory.)

Much closer in spirit to a modern structuralist view is the earlier attitude of Dedekind in his classic essay, 'Was sind und was sollen die Zahlen'.[5] There, having provided an analysis of what we today call an ω-sequence—in Dedekind's terminology, a "simply infinite system"—as any system of elements together with a relation (of "successor") satisfying four axioms (in which we recognize the "Peano postulates", with a second order statement of induction), Dedekind made the following remarks under the heading "Definition":

If in the consideration of a simply infinite system N set in order by a transformation ϕ we entirely neglect the special character of the elements; simply retaining their distinguishability and taking into account only the relations to one another in which they are placed by the order-setting transformation ϕ, then are these elements called *natural numbers* or *ordinal numbers* or simply *numbers*, and the base-element 1 is called the *base-number* of the *number-series* N. With reference to this freeing the elements from every other content (abstraction) we are justified in calling numbers a free creation of the human mind. The relations or laws which are derived entirely from the conditions $\alpha, \beta, \gamma, \delta$ in (71) [Dedekind's statement of his "Peano axioms"] and therefore are always the same in all ordered simply infinite systems, whatever names may happen to be given to the individual elements, ... form the first object of the *science of numbers* or *arithmetic*.

Later in the essay, having proved in effect that any two simply infinite systems are isomorphic (Theorem of §132) and a kind of converse, that any system isomorphic to a simply infinite system is also simply infinite (Theorem of §133), Dedekind remarked (§134):

... it is clear that every theorem regarding numbers, i.e. regarding the elements n of the simply infinite system N set in order by the transformation ϕ, and indeed every theorem in which we leave entirely out of consideration the special character of the elements n and discuss only such notions as arise from the arrangement ϕ, possesses perfectly general validity for every other simply infinite system Ω set in order by a transformation ψ ... By these remarks, as I believe, the definition of the notion of numbers given in (73) is fully justified.

Now, it is quite clear from these passages that Dedekind endorsed a structuralist view to this extent: he regarded arithmetic as fundamentally the study of *relationships within arbitrary ω-sequences* and regarded the "special character" of the items of any such sequence as of no concern whatever to mathematics. Surely this is why he took pains to prove the categoricity of his axioms and the converse (which, in a modern model-theoretic treatment we would fill out with a proof that isomorphic models are *elementarily equivalent*, i.e. agree on all the same sentences, not just the axioms).

Yet can we say that Dedekind endorsed the *further* structuralist thesis that, to make sense of arithmetic, it is not necessary to regard number words as referring to any objects whatever, and that it is certainly not necessary to construe them as referring to unique abstract objects, called "numbers"? It is tempting to read Dedekind as going this far, but, apparently, that would be to *misread* him.[6] It is tempting to read 'N' and 'ϕ' in §73 quoted above as (universally quantified) variables, typical of definitions. However, as emerges in subsequent sections of the essay—including §134 which we have quoted—Dedekind in fact is using 'N' and 'ϕ' as constants, as names for a particular system of abstract objects and a particular relation on that system, respectively. (Note the definite article as it occurs in §134. And note that the very statement of the theorem of §132—"All simply infinite systems are similar to the number-series N and consequently by (33) to one another"—becomes quite redundant on the universal quantifier reading.) These objects are indeed the unique referents of number words, "the natural numbers". They are abstract particulars; and they are created by our minds (the very same ones by each of us?); and, presumably, they lack any "special character" (even the character of having been created by Richard Dedekind, or by *homo sapiens sapiens*, as he likes to call himself?). Whatever the precise position—and regardless of whether it is coherent—

---

[6] See the references cited above, n. 11 to the Introduction, esp. Parsons [forthcoming].

Dedekind's seems to fall into the objects-platonist category after all. And, it seems, with a vengeance: not only is the position platonist; it is creationist as well!

Of course, we need not follow Dedekind in all of this (if indeed this rather literal reading of his words is accurate). We can attempt to retain his structuralist insights and strip away his version of platonism. One way of doing this would be simply to adopt a modern set-theoretic treatment of number theory, supplemented with an explicit recognition that number theory isolates an isomorphism type, a class of pairwise isomorphic structures, any one of which can serve as providing "referents" of number words. (A detailed analysis of natural language "reference to numbers" could be developed in various ways, which need not detain us here.)

However, as already indicated in the Introduction, we seek an alternative to the set-theoretic articulation of structuralism. Not only does that path involve the categorical assertion of set-theoretic axioms with their "abstract commitments" (pending a structuralist reinterpretation of set theory itself, something which is fraught with special problems, as we shall see in the next chapter); it makes number theory dependent on set theory in a way that, from a *mathematical* point of view, it would be desirable to avoid. There is good motivation for understanding number theory and analysis as capable of standing on their own. Surely we should resist saddling them—as basic mathematical theories—with the problem associated with "Cantor's universe" (i.e. with "unsolvable problems the likes of which mathematics has never known before").[7] Indeed, it is possible to give set-theoretic "reductions" of these more basic theories in weak systems of set theory. But then, in addition to the categorical commitments to sets, sets of sets, sets of sets of sets, etc. (at least through all finite levels), there is the problem that our structuralism for the more basic theories will turn out to be *too restricted*: restricted, that is, to whatever structures of the appropriate type can be found within the weak set-theoretic framework (which, of course, may include urelements). What will be missed is the full *generality* of structuralism: arithmetic or analysis investigates relations holding within *arbitrary* structures of the appropriate type—not just within those that happen to be recognized in a weak set theory.

---

[7] Barwise [1977], p. 41.

As an alternative, we shall make limited use of a logico-mathematical modality—a notion of logical possibility—as part of the structuralist language. Just how this is to be understood will emerge from the role it plays in the interpretation. By employing it, as we shall see, it will be possible to translate ordinary sentences of number theory (or analysis) so that, on the interpretation, they say what *would* be the case in any (arbitrary) structure of the appropriate type without literally quantifying over any objects at all. (For definiteness, we carry out the analysis for Peano Arithmetic in the first instance. It can readily be adapted to more powerful theories, such as Real Analysis, as will be made clear later.) This, of course, must be spelled out by means of a specific translational pattern, and that will be presented below (§ 1). Since modal conditionals are involved, I call this the *hypothetical component* of the (ms) interpretation ("msi"). However, as will emerge, there must also be a *categorical component*, a set of basic assumptions of the structuralist theory, from which the mathematics as ordinarily practised can be recovered. As we shall see, care in the choice of postulates will allow a natural derivation of the (first) *axiom of infinity* (in the relevant form of modal existence) (§ 2). The use of modality will thus have mathematical foundational relevance beyond the more "philosophical" questions of "platonism". Once these components have been set out, we can turn to the questions of justification. These are some of the most interesting problems confronting the approach. As we shall see (§ 3), it is relatively easy for the set theorist (platonist) to justify the translation scheme (and the categorical component as well), but this is a justification "from the outside". If the structuralist view is to stand on its own, it ought to be able to answer certain questions concerning truth-determinateness *from the inside*, i.e. without going outside the ms framework. Once the translational pattern has been given, it will become apparent what must be established, and we will outline the key steps in what seems to be a satisfactory solution (§ 4). It will then be indicated how these methods can be extended to a ms treatment of more powerful theories, in particular Real Analysis, including the second-order version involving quantification over sets of reals (§ 5). As a kind of corollary, it will emerge that a great deal of classical mathematics—far more than generally supposed—can be understood "nominalistically" (apart from the modality); and the internal justification can be so understood as well (§ 6).

§ 1. The Modal-Structuralist Framework: The Hypothetical Component

Intuitively, we should like to construe a (pure) number-theoretic statement as elliptical for a statement as to what would be the case in any structure of the appropriate type. In this case, the structures are, of course, "progressions" or "ω-sequences", so what we seek to make precise is a translation pattern that sends a sentence of arithmetic S to a conditional such as,

If X were any ω-sequence, S would hold in X.   (1.1)

We now confront two distinct problems in formalizing this, corresponding to two distinct desiderata of the programme: (i) to avoid literal quantification over abstract structures, possible worlds, or intensions, in order to provide a genuine alternative to objects-platonism, in which literal reference to such objects is eliminated; (ii) to respect the full, classical truth-determinateness of the mathematical theory in question. (i) guides us in the matter of relative placement of modal operators and quantifiers while (ii), on the other hand, constrains the choice of non-modal primitives to be admitted into the modal-structuralist language. Let us consider these in turn.

Note that (1.1) contains, implicitly, a universal quantifier. It is already a step toward formalizing the more purely English version,

If there were any ω-sequence, S would hold in it,   (1.1′)

in which the pronomial cross-reference indicates that the apparent existential quantifier is really universal (on the plan of ordinary examples, such as "if someone objects to this procedure, she or he should speak up now", so confusing to beginners in logic). However, the "if there were" serves to distinguish the intended sense from that of "For any (actual) x, if x were ...", where we contemplate of each (actual) thing what would be the case were the antecedent to hold of it. (Think of simple examples, such as, "if there were seven-legged horses ...", as opposed to, "If anything (i.e. any existing thing) were a seven-legged horse ...".) The upshot of this is that we should represent (1.1) as having the outer logical form,

$\Box\forall X(X$ is an ω-sequence $\supset$ S holds in $X)$,   (1.1″)

with the quantifier within the scope of the modal operator. So long as we adhere to this plan, our translates will have the intended generality. Moreover, we will attempt to formulate existential assumptions in a parallel fashion. The categorical component of the interpretation (of PA) will assert

$$\Diamond \exists X (X \text{ is an } \omega\text{-sequence}), \tag{1.2}$$

not

$$\exists X \Diamond (X \text{ is an } \omega\text{-sequence}).$$

Thus, immediately we see that the background modal logic must not contain the Barcan formula. (As to what else it should contain, the natural choice will be S-5 since we are concerned here with an absolute, mathematico-logical sense of possibility.)[8] As a result, the interpretation will be neutral with respect to the actual existence of mathematical structures. And as a consequence of this, questions of actual reference to such structures as objects will simply not arise. The structuralist intuition—that such questions as, "How do our words attach to abstract structures in the absence of any interaction between them and us?" are misguided—will be sustained.

It should also be stressed that, in our unwillingness to quantify over possibilia, we avoid such extravagances as "the totality of all possible ω-sequences". From our point of view, such totalities are illegitimate, much as "the totality of all possible sets" is illegitimate. While it may be possible to treat such totalities without literal contradiction, countenancing them runs counter to the open-ended

---

[8] A further reason for this choice is that we never seem to require iterated modalities to express what we need to express. (In S-5, all modalities are reducible to those of a basic short list involving no iteration.) Moreover, the choice of S-5 leads to the pleasant result that widely shared intuitions concerning the "collapse of modalities" in connection with mathematical objects are actually sustained on translation: where 'O' is a predicate for such objects, the statement e.g. that

If Os are possible, then Os are actual (i.e. actually exist)

can be understood as asserting that

$$\Diamond \Diamond \exists x(O(x)) \supset \Diamond \exists x(O(x)),$$

where "actual (mathematical) existence" is interpreted as (mathematical) possibility; and the statement that

If Os are actual, then Os are necessary (i.e. necessarily exist),

comes out as

$$\Diamond \exists x(O(x)) \supset \Box \Diamond \exists x(O(x)).$$

The first formula is an instance of the characteristic S-4 modal axiom; the second an instance of the characteristic S-5 modal axiom (which implies the S-4 axiom, incidentally).

character of mathematical construction, which the msi seeks to respect. (In the present case, any totality of ω-sequences could form the basis for a new one, much as any totality of sets satisfying, say, the ZF axioms could be extended to a richer model. Such "extendability principles" are taken up in Chapter 2.)

So much by way of outline on the first problem of formalization. What about the second? Here we need to express the notion of "ω-sequence" and, it seems, "holding" or "satisfaction". Now one way of accomplishing this would be simply to use the language of set theory, since we know how to express both "ω-sequence" and "satisfies" in terms of set membership; (1.1) could then be made precise by

$$\Box \forall X (X \vDash \wedge PA^2 \supset X \vDash S), \tag{1.3}$$

where $\wedge PA^2$ is the conjunction of the (finitely many) second-order Peano axioms, and satisfaction is defined so as to guarantee that models of these axioms are "full", i.e. the second-order quantifier in the induction axiom,

$$\forall P[\{\forall x(\forall y(x \neq s(y)) \supset P(x)) \,\&\, \\ \forall n(P(n) \supset P(s(n)))\} \supset \forall n P(n)],$$

ranges over all subsets of the domain of X.[9] One disadvantage of this choice is that the translates all become metalinguistic, and this is surely an awkwardness, if not a fatal misrepresentation of arithmetic discourse. But even more serious is the problem that the structuralist programme, so articulated, becomes just a piece of modal set theory, and for reasons already stated above, we do not wish to follow this course.

Nevertheless, if we are to insure the full truth-determinateness of our translates, *some* higher-order notions will be needed. Even if we could write out all the first-order axioms of PA (i.e. with all instances of the induction scheme), taking this infinite conjunction as the antecedent of a modal conditional would not express the intended restriction to ω-sequences (as a familiar Henkin compactness argument shows). But unless such a restriction is expressed, the scheme will confront problems in connection with non-standard models: sentences not decided by the first-order axioms will all be treated as "not true", in that they would not hold in all models of the

---

[9] NB. X here, of course, is a pair, a domain together with a valuation interpreting the predicate (or function) constants of $\mathscr{L}(PA^2)$. A single such, for successor, suffices.

axioms. The structuralist interpretation would then lapse into a semantic equivalent of deductivism.[10]

The point bears emphasis, and can be brought out by considering the categorical modal existence assumption, (1.2), that, as we shall insist, must accompany the hypothetical component of any adequate modal-structural interpretation. The intended assumption, intuitively, is

ω-sequences are possible,

not the weaker claim,

Some model of the first-order axioms is possible.

The latter is equivalent to the mere claim of formal consistency, whereas (1.4) expresses more. In familiar parlance, (1.4) expresses the possibility of a *standard* model, not an arbitrary model. But it is the stronger claim, not the purely formal claim, that a realist structuralism requires. (This point will emerge in a more precise form below, when we turn to questions of justifying the translation scheme. For now, simply note that, once we interpret arithmetical sentences as elliptical for what would hold in a class of (hypothetical) structures, such translates will be true only if all the structures in that class agree on the sentence in question. Immediately, we see that all the relevant structures must at least be elementarily equivalent. The structuralist naturally desires that they moreover be pairwise isomorphic. As is well known, the models of any first-order theory of arithmetic—even the highly non-recursively-enumerable theory of all truths of the standard model—do not satisfy this latter condition.)

Thus, the modal-structuralist framework for arithmetic must transcend first-order arithmetic, even in its non-modal primitives. And, as just indicated, here "for arithmetic" means "for either first-order or second-order arithmetical sentences". (That is, even if we restrict the translation to first-order sentences, higher order primitives will be needed to insure truth-determinateness of the scheme.) If, however, we wish to stop short of full set theory, we are naturally led to consider languages which make some, but only a modest, use of

---

[10] This problem is recognized in Kessler [1978].

set-theoretical or allied concepts. Our prime candidate here is the framework of second-order logic.[11]

Now, as Parsons has recently put it, one of the beauties of second-order logic is that it is susceptible of so many different interpretations.[12] Interpretation there must be, for, as Quine has insisted, some predication relation is hidden in the notation 'F(x)', once we admit F as a quantifiable variable. And if we eschew intensional objects (attributes), we are naturally led, in the first instance, to construe F as an extension, i.e. as a class, in which case predication is indeed (the converse of) membership. However, to countenance classes of individuals is one thing; to countenance iteration of the power set operation is another (not to mention iteration into the transfinite!).[13] If we abstract from the concerns of a strict nominalism, this extensional interpretation of second-order logical notation is a natural candidate for articulating structuralism. In effect, we allow classes as occurring exclusively on the right of ∈: we collect whatever individuals of a domain we are given (actual, or hypothetical), but we go no further in "collecting these collections". Similarly, we allow talk of arbitrary k-place relations (hence functions) on a given domain of individuals—relations obeying extensionality—but we do not go further and collect or relate these. (Formally, we shall adopt the ordinary comprehension axioms of second-order logic. See below, Chapter 1, §2.) Not that it would be incoherent to do so. Rather, to do so is to introduce iteration, hence a theory of iteration, i.e. set theory, and, on the structuralist view, this is a special mathematical theory investigating structures of its own. How, ultimately it should be treated remains to be determined, but, as already indicated, we wish to give a structural interpretation of theories "lower down" which is independent of the full-blown iterative theory. To do so, we need to speak, at least hypothetically, of structures, and the

---

[11] In addition to its naturalness as a framework for formulating structuralist principles, the choice of second-order logic is also motivated by the consideration that we should like to leave open the possibility of a "nominalist" reading of the mathematical theories in question. As we shall see below (§6), such a reading is indeed possible. It would not be, however, were our framework third or higher order.

Nominalism apart, second-order logic forms an interesting alternative to set theory as a background for formulating various mathematical theories. For an especially insightful recent example, see Boolos's reconstruction of Frege's *Die Grundlagen der Arithmetik*, cited above, p. 11, n. 3.

[12] See Parsons [forthcoming].

[13] Cf. Shapiro's [1985] distinction between "logical" and "iterative" conception of "set". Cf. also Maddy [1983] and Tait [forthcoming].

language of classes and relations embodied in second-order logic is well suited to the task.

Two related final preliminaries: first, in employing the phrase "second-order logic", we are referring to a well-known notation and its metatheory; we are not committed to the view that it is "genuine logic". Nor are we committed to *any* particular way of drawing a line between logic and mathematics. As we see it, structuralism does not need to draw such a line. What matters here is not whether second-order logic is logic, but whether it is intelligible and useful. As the remarks above indicate, we do see a certain mathematical content in the second-order formalism, and it is precisely this content that is to be exploited in articulating structuralism. In fact, as will emerge in a moment, the non-schematic character of second-order logic—i.e. the possibility of forming fully quantified sentences which, intuitively, have truth value—turns out to be a crucial advantage for the structuralist programme.

Second, we are exploiting the well-known *superior expressive power* of second-order axiomatizations of mathematical theories, manifest primarily in the fact that, by means of finitely many second-order axioms, a single type of structure can be characterized up to isomorphism.[14] The significance of categoricity has been a subject of controversy.[15] In part, we believe this to stem from failure sufficiently to distinguish two very different functions of axiom systems. On the one hand, there is the purpose of codifying proofs; here, what we usually demand of proofs leads us to work with first-order systems. But, on the other hand, there is the very important, distinct purpose of "expressing what we mean"—on a structuralist view, the purpose of articulating the type of structure of mathematical interest.[16] A good deal of mathematical work itself may go into achieving this, and, of course, it would be anachronistic to insist that, prior to codification, mathematicians "didn't know what they were talking about". But understanding comes in degrees, and once a categorical articulation has been achieved, it can be used to communicate more definitely what is intended. Moreover, it can be used in developing an interpretation aimed at philosophical clarification.

Now, once we have the apparatus of second-order logic, we can

[14] This has been emphasized by many authors. See e.g. Montague [1965], Kreisel [1967], Shapiro [1985]. Interestingly, the issue is not considered in Dummett [1963].

[15] See e.g. Weston [1976], Corcoran [1980], Shapiro [1985].

[16] We are in agreement with Shapiro [1985], where this distinction is also drawn.

attempt to express the hypothetical component of the msi in a direct, mathematical—as opposed to metamathematical—way. In the case of PA, we can write out the conjunction of the finitely many PA^2 axioms, abbreviated '$\wedge PA^2$'.[17] It is tempting then to take S in $\mathscr{L}(PA^2)$ simply as elliptical for

$$\Box(\wedge PA^2 \supset S). \tag{1.5}$$

This has the virtue of simplicity, but I think the msi must do better. For, as it stands, (1.5) is problematic. It contains, in addition to purely logical notation, at least a relation constant, 's', for successor. But, if (1.5) is thought of as making a definite assertion, how are we to interpret such constants? What, for example, does it mean to say, "If there were an object which were *the successor* of no object & ..." (which is one of the PA^2 axioms)? The platonist claims to understand "successor" as a particular relation on a particular domain— either the natural numbers, unreduced, or identified (arbitrarily) with a particular progression of sets, in which case the successor relation is set-theoretically defined. But the structuralist wishes to depart from both styles of platonist interpretation and substitute something in a way more abstract, which we wrote, intuitively, as (1.1) above. (1.5) simply does not articulate (1.1).

If the term 's' is not understood as having a definite extension, how is it to be understood? The most likely answer, from a modalist, might be "as expressing an intension", e.g. as a distinct function in distinct possible worlds. But, in this case, one has ascended to a possible worlds semantics of $\Box$, and the ms framework is really an objects-platonist one after all. (Alternatively, one could simply take intensions as primitive, but this is surely no way to avoid objects-platonism.) In effect, what has happened is that the predicate constants have, in the modal contexts, ceased to function as genuine constants (in the usual sense, of having definite extensions). They are behaving rather as predicate variables. But, then we have in (1.5) not a sentence but a scheme. In this case, we would be tempted to ascend semantically in order to obtain definite sentences. We might, for example, introduce appropriate semantical machinery with which we could then express that (1.5) is a modal-logical truth. But then (a) our translate of S is really this metamathematical statement, and not (1.5); and, more important, (b) the ms framework has embraced

---

[17] For a clear presentation of PA^2 and of axiomatic second-order logic, see Robbin [1969], ch. 6.

possible worlds or a full-fledged model theory after all, contrary to its aims.[18]

Now one advantage of the second-order framework is that we can bypass these problems by quantifying over relations. We can form, e.g., the sentence,

$$\Box \forall f (\wedge PA^2 \supset A)\binom{s}{f}, \tag{1.6}$$

in which a two-place relation variable f replaces the constant s throughout the conditional. This is now a direct, modal-mathematical sentence corresponding to the more familiar metamathematical claim that A is logically implied by the PA^2 axioms (in the sense of full second-order logical implication, i.e. A holds in all full second-order models of $\wedge PA^2$).

While (1.6) improves significantly upon (1.5), there is still a respect in which it is schematic. This concerns the question of domains of the (possible) structures, explicitly involved in our initial intuitive formulations (e.g. (1.1)), but about which there is no mention in (1.6). Usually, this would be brought in at the level of metatheory, as in the talk of logical implication just cited. But we do not want to rely on metatheory to express even this component of structuralism. There is, however, a natural remedy: we can employ the device of relativized quantification with respect to a class variable X, and then prefix the formula (following the $\Box$) with a universal quantifier, $\forall X$. This gives us

$$\Box \forall X \forall f [\wedge PA^2 \supset A]^X \binom{s}{f} \tag{1.7}$$

as our direct, modal mathematical translate, A_{msi}, of A of $\mathscr{L}(PA^2)$.[19]

---

[18] Something very much like this predicament was encountered by Kessler [1978] in his critique of Putnam [1967], where the modal translates were taken to be first-order modal-logical schemata. The relevant distinctions it is necessary to make among these e.g. between those that are valid and those that are not, then require a possible worlds semantics, or something equally powerful.

[19] First-order quantifiers inside the bracketed portion are relativized in the usual way, i.e.

$$\forall x \phi \to \forall x (X(x) \supset \phi), \quad \exists x \phi \to \exists x (X(x) \& \phi),$$

and higher-order quantifiers $\forall R, \exists R$, are relativized by employing formulas asserting that the relata of R are 'in X', i.e. for k-ary R:

$$\forall R \phi \to \forall R (\forall x_1 \ldots \forall x_k (R(x_1 \ldots x_k) \supset X(x_1) \& \ldots \& X(x_k)) \supset \phi),$$

$$\exists R \phi \to \exists R (\forall x_1 \ldots \forall x_k (R(x_1 \ldots x_k) \supset X(x_1) \& \ldots \& X(x_k)) \& \phi).$$

If we wish to restrict attention to translates of $\mathscr{L}(PA^1)$, we follow the same pattern, but with the following changes: since the original language has symbols for addition and multiplication (say Σ and Π), either we may define these operations explicitly from successor and second-order quantification in the familiar way, or we may add the recursion equations for these as axioms to PA^2 (obtaining PA^{2+}); then, when representing a sentence A of $\mathscr{L}(PA^1)$, we replace Σ and Π respectively throughout with three-place relation variables, so that A_{msi} becomes,

$$\Box\forall X\forall f\forall g\forall h[\ \wedge PA^{2+} \supset A]^x\binom{s,\ \Sigma,\ \Pi}{f,\ g,\ h}. \tag{1.8}$$

This, expresses—in effect—that A would hold over any domain and with respect to any relations on that domain "behaving as successor, addition, and multiplication". The part about behaviour is guaranteed by the axioms of the antecedent; but "A would hold" is obtained, without introducing satisfaction, by means of the relativized form of A with the substitutions indicated. (Saying "A would hold" is misleading in that, of course, the substitutions of 'f' for 's', etc., are made throughout A as well as throughout the axioms.)

Let us adopt the schemes, (1.7) and (1.8), as giving the hypothetical component of the modal-structuralist interpretation of $\mathscr{L}(PA^2)$ and $\mathscr{L}(PA^1)$, respectively. So far, they do nothing more than represent explicitly the "modal-structural content" of ordinary sentences of the respective languages. But, of course, there is much more to mathematical practice than the mere utterance of sentences. There is the deductive practice of theorem proving, and for the classicist, there are claims of truth and falsity accompanying that practice. Let us see how, in broad outline, the modal-structuralist can understand these fundamental matters.

§ 2. The Categorical Component: An Axiom of Infinity and a Derivation (inspired by Dedekind, with help from Frege)

To begin, it should be clear how the practice of theorem proving can be "recovered". If we make the simplifying assumption that that practice is already reasonably codified by standard axiom systems for PA (to fix ideas, let us say first-order PA), then the modal-structuralist has a straightforward way of arriving at T_{msi}, for any theorem T of PA. The first step is to adopt a standard axiom system of second-order logic, including the full comprehension scheme,

$$\exists R \forall x_1 \ldots \forall x_k [R(x_1 \ldots x_k) \equiv A], \tag{CS}$$

where the x_i are individual variables, R is not free in A, and A may contain parameters, but no modal operators. (This latter restriction enables us to avoid quantification into modal contexts at this stage of the programme.) (CS) will enable one to derive, among other things, any instance of the first-order induction scheme from the second-order induction axiom. The next step is to apply the translation scheme ((1.8), in this case) to each line of the original PA proof. The axioms of the original proof go over to (necessitations of) theorems of second-order logic, and it is a (tedious) exercise to show that the rules of inference preserve this relation. (One needs to show that the relativization of a theorem is a theorem.) Thus, if T was correctly deduced from (finitely many) first-order axioms, T_{msi} will be deduced in second-order logic together with the most elementary rules for $\square$.

It should be noted that, in view of the rule of necessitation of axiomatic modal logic (in particular, the system S-5, to which we shall appeal), we have, not merely (CS), but its necessitation,

$$\square \exists R \forall x_1 \ldots \forall x_k [R(x_1 \ldots x_k) \equiv A], \tag{$\square$CS}$$

with the same restrictions given for (CS) above. This will prove useful below, at several points in the justification of the interpretation. Observe that the initial $\square$ is the only modal operator occurring in these comprehension principles. (The importance of this will emerge below, cf. p. 31). Also note that the restriction that the x_i range over individuals—a restriction built into axiomatic second-order logic to block paradox (e.g. Russell's)—accords with our above-mentioned aim of avoiding the iteration of any collecting (or relating) operation: with ($\square$CS) we recognize (at most) only one abstract level, not a complex hierarchy of classes and relations.

Returning to the matter of recovering proofs, a short cut runs as follows: replace all relation constants in the original proof of T with relation variables of the appropriate type; use the deduction theorem to justify conditionalization, inferring,

$$[\wedge AX \supset T]\binom{s, \Sigma, \Pi}{f, g, h},$$

where $\wedge \mathrm{A}X$ is the conjunction of the axioms used to derive T; replace the antecedent with $\wedge \mathrm{PA}^2$ (justified by the fact that each instance of the induction scheme is derivable from the second-order induction axiom using (unrestricted) second-order comprehension); relativize the quantifiers to X; apply universal generalization to the second-order variables, and necessitation to this. Thus, ordinary proofs are simply construed as free variable arguments (the variables being second order) relative to an arbitrary domain. In this way, deductive practice (codified) in the original system is preserved.

Obviously, it is not being recommended that ordinary proofs be carried out in this way (any more than the proof theorist recommends that ordinary proofs be replaced in practice by any of the usual codifications!). The point is merely that the translation scheme is proof-theoretically faithful in that, in principle, ordinary theorems can be recovered in the appropriate form and within a framework in which apparent reference to numbers or to "the standard model" as an object has been entirely eliminated.

Recovery of proofs is, however, only the first step in justifying the translation scheme. As already emphasized, the modal-structuralist aims at much more: in some suitable sense, the translates must be mathematically equivalent to their originals, where these are understood classically, i.e. as determinately true or false (or true or false in the standard model). Just how this "equivalence" is to be understood and how it may be established are questions to be addressed in the following sections. Here a first step may be taken in that direction by drawing attention to what we have already called the "categorical component".

As already suggested, a categorical assumption to the effect that "ω-sequences are possible" is indispensable and of fundamental importance. Without it, we would have a species of "if-thenism", i.e. a modal if-thenism, and this would be open to quite decisive objections, analogous to those which can be brought against a naïve, non-modal if-then interpretation. Consider the latter. Suppose it represents sentences A of arithmetic by means of a material conditional, say, of the form,

$$\wedge \mathrm{PA}^2 \supset \mathrm{A},$$

or some refinement thereof. Suppose also that, *in fact*, there happen to be no actual ω-sequences, i.e. that the antecedent of these conditionals is false. (This could be "by accident" as it were. For the sake of argument, do not insist upon Cantor's universe of sets as "necessary existents" (please!). Consider, instead, the stance of the "if-thenist" who seeks to avoid platonism.) Then, automatically, the translate of every sentence A of the original language is counted as true, and the scheme must be rejected as wildly inaccurate. (Well, at least it gets half the answers right—not the worst imaginable performance! Compare the case of the broken watch.)

Now the very same situation would obtain in the case of modal conditionals if ω-sequences were *not possible*, i.e. if there could (logically) be no standard realization of the PA^2 axioms. (This could be due to a formal contradiction, but, as we know from Gödel's work, absence of a formal contradiction is insufficient for the possibility of a standard model. For example, suppose there were an ω-inconsistency.) In that case, the translation scheme would not respect negation: all the original sentences A would be translated as true. Thus, it is absolutely essential to affirm, categorically, an appropriate version of (1.4), above. In the style of our preferred version of the hypotheticals, we can adopt,

$$\Diamond \exists X \exists f (\wedge PA^2)^X \binom{s}{f},\tag{1.9}$$

as a basic thesis of modal mathematics. On the view under consideration, this affirms the coherence of the notion of an ω-sequence, something that is generally taken for granted, but which nevertheless forms an indispensable "working hypothesis" underlying mathematical practice.

In fact, on the modal-structuralist "rational reconstruction" of (pure) number-theoretic practice, the modal-existence postulate, (1.9), can be viewed as a starting-point in ordinary reasoning "about numbers". As we have just seen, that reasoning *can* be reproduced entirely in terms of the ms-translates, in which all apparent reference to numbers as objects disappears. But this is a cumbersome substitute for ordinary reasoning, which appears to be "about" a fixed ω-sequence. Now, given a categorical assumption such as (1.9), the ordinary reasoning can be seen as a natural short-cut, and the use of number-words as apparently referring constants receives a natural interpretation. One begins with (1.9), informally put: "Assume an ω-sequence".[20] Next one applies "existential instantiation", i.e. one

[20] In the course of criticizing the "cornucopian economist's" approach to ecological problems, Paul Ehrlich recently related the following anecdote: He and an economist friend (not of the cornucopian school) were on an outing and couldn't find

says, let N (i.e. (N, s)) be such a thing. Formally, this amounts to writing $(\wedge PA^2)^N$. Here 'N' is just a "dummy name", serving to facilitate our reasoning in the usual ways. One doesn't claim that 'N' refers to anything, any more than one claims that 'a' refers to anything when one passes from a non-modal existential assumption— e.g. "there are black holes", to "let a be a black hole". But, now, one will not hesitate to introduce the individual numerals as a convenience, justified by the existence and uniqueness claims derivable from $(\wedge PA^2)^N$, i.e. one will introduce '0' as "standing for the first member of N", '1' as "standing for the unique next member of N", and so on. All further arithmetical reasoning will be carried out with respect to this fixed N, and it will appear that there is constant reference to objects. But all of that is just a *façon de parler*, justified as a short cut, beginning with a mere claim of logical possibility.

At this point, it is worth comparing the procedure just outlined with the "abstraction" that Dedekind employed in his "definition" of *natural number*, quoted at the outset. For, without claiming that Dedekind would have endorsed the msi, I think we can see in the above recovery of ordinary number talk a way of sustaining his structuralist insights without lapsing into objects-platonism or creationism. Like Dedekind, we appeal to a second-order definition (essentially, Dedekind's) of the type of structure of interest. But, instead of "creating abstract particulars", we appeal to the logical possibility of structures of the relevant type. Moreover, in entertaining such structures and in mathematical reasoning concerning them, we "neglect entirely the special character of the elements, simply retaining their distinguishability and taking into account only the relations to one another in which they are placed" by the hypothesized order relation of the structure. (Cf. §73 of Dedekind [1901] quoted above.) Furthermore, we do even arrive at talk of numbers as objects, but this is just a convenience justified by our assumptions. Finally, creation even enters as well. However, it is not, of course, "the objects" that we create, but the *language of numerals*—again, justified by our assumptions, and by the need for efficient, compact forms of reasoning.

In one major respect, however, our procedure thus far departs radically from Dedekind's. For, whereas we have been taking the

---

a can-opener. "Why don't we adopt the economist's solution to this?" the friend said. "What's that?" Ehrlich asked. "Assume a can-opener!" he replied. Mathematics extends cornucopian economics into the transfinite.

modal existence of ω-sequences (1.9) as a postulate, Dedekind attempted to *derive* the existence of simply infinite systems from more basic considerations.[21] And it may well appear that, at this point, we must part company with Dedekind. For recall that there were at least two flaws in Dedekind's "proof": first, there was the problem of meaningfully iterating a "the thought that ..." operator an arbitrary finite number of times, obtaining a new object at each stage. And, second, there was the need to collect all such objects via some comprehension principle, which Dedekind did not explicitly articulate. Frege, of course, did explicitly articulate such a principle, but it went too far. As with Russell, modern set theory simply adopts an axiom of infinity without seeking a derivation, and it is easy to acquiesce in such a policy. (As Russell stressed, the existence of infinitely many (non-mathematical) objects is a *contingent* matter on which mathematics should not depend. And, once we accept cumulative set theory, *à la* Zermelo, restricted comprehension (*Aussonderung*) no longer enables us to generate the set of all (set-theoretically constructed) natural numbers without already having an embracing set. The axiom of infinity seems to be a true "axiom".)

It is worth pointing out, however, that in the present framework—axiomatic second-order modal (S-5) logic—a natural derivation of the mathematical existence of infinite sets, and indeed, of ω-sequences, can be carried out, along quasi-constructive lines.

One begins, in the manner of Dedekind, with a rule prescribing the construction of a unique "next object". It could, for example, prescribe that a new stroke be added at the end of any presented terminating sequence of strokes (in some fixed orientation). Call the rule R. Let $A(x, y)$ be the predicate, "y is generated after x in accordance with rule R". Now, using this predicate, we can write down a first-order sentence saying, in effect, that the field of the relation A is infinite:

$$\exists x \exists y (A(x, y)) \ \& \ \text{"A is asymmetric and transitive"} \ \&$$

$$\forall x \exists y (A(x, y)) \ \& \ \forall x \exists! y (A(x, y)) \ \& \ \sim \exists z (A(x, z) \ \& \ A(z, y)). \qquad (*)$$

(Here, it is assumed that the quantifiers are restricted to objects of the appropriate sort (strokes); this can be absorbed by the predicate

---

[21] Dedekind [1901], §66, which contains an abortive "proof" that there exist infinite systems, based on the idea that reflection on one's thoughts yields a new thought. (From this the existence of simply infinite systems would follow. Cf. §72.) We comment on this briefly below.

A. The discreteness condition expressed in the last clause is added to facilitate the connection with ω-sequences, to be brought out momentarily.) Now, in fact, there may be no reason to accept (*), but there is every reason to accept that, logically, it might be true,

$$\Diamond(*). \qquad\qquad \text{(Potential Infinity)}$$

Now, appealing to comprehension (CS), the following is provable:

$$\Box[(*) \supset \exists X \forall z \{ (X(z) \equiv \exists x A(x, z))$$

& "X includes an ω-sequence"$\}$],

where the quoted clause is spelled out in second-order notation as in (1.9). Then, appealing to modal logic and the assumption $\Diamond(*)$, we obtain

$$\Diamond \exists X \forall z [(X(z) \equiv \exists x A(x, z)) \ \& \ \text{"}X \text{ includes an } \omega\text{-sequence"}],$$
$$\text{(Modal Infinity)}$$

and, then, of course,

$$\Diamond \exists X \exists f (\text{PA}^2)^X \binom{s}{f}, \qquad\qquad (1.9)$$

our key modal-existence thesis for arithmetic.[22]

Two remarks are in order concerning this derivation. Note, first of all, that it employs (CS), *ordinary* second-order comprehension (CS). (Equivalently, if one works, say, with semantic tableaux, one can begin with $\Diamond(*)$ and then invoke ($\Box$CS), above.) Only ordinary (actualist) universal quantifiers are involved. The derivation does *not* employ "modal comprehension", either in the sense of comprehen-

---

[22] It should be noted that "potential infinity" involves accepting, not only $\Box \forall x \Diamond \exists y A(x, y)$, corresponding to a motion from any world with a "stroke" to a (possibly different) world with a successor, but the stronger claim that $\Diamond \forall x \exists y A(x, y)$, i.e. in a single world each stroke has a successor. (This move corresponds to postulating the "union" of the domains of strokes of the worlds answering to $\Box \forall x \Diamond \exists y A(x, y)$, something it is always possible to resist without literal contradiction.) But Euclidean time or space models this kind of condition; and even arch-opponents of "completed infinities" concede the coherence of such models. Cf. e.g. Dummett [1977], who grants that "it is perfectly intelligible, even if in fact false, to say that there are infinitely many stars" (p. 57). This concession is, however, regarded as irrelevant to "mathematical totalities, whose elements are mental constructions" (p. 58). (Yet against the objects-platonist's retort that mathematical objects are eternally existing abstract objects, it is said that "the question is not, however, resolved by the mere utterance of a meta-physical credo"! (ibid.).) From the modal-structuralist point of view, this debate can be bypassed entirely. Dummett has conceded all that the classical mathematician requires of a first-level infinite totality.

sion for formulas with modal operators, or in the sense of a scheme with "possibilist quantifiers", of the form,

$$\exists R \Box \forall x_1 \ldots \Box \forall x_k [R(x_1 \ldots x_k) \equiv A].$$

It must be stressed that this latter is no part of the modal-structuralist framework we are investigating. It would give rise to intensions, and moreover to monstrous intensions such as "the union of all possible ω-sequences", which, as already indicated, runs counter to the spirit of the interpretation. (This is closely connected with the msi's rejection of proper classes. We shall return to this in Chapter 2.) The version of comprehension actually employed simply allows ordinary mathematical reasoning to proceed under hypotheses entertained counterfactually (or, more accurately, "neutrally" with respect to actual existence).

Second, note that the derivation is similar to Dedekind's attempt, in that it *uses* a predicate which we suppose we understand, and, moreover, a predicate pertaining to a particular sort of object of no special concern to the mathematical theory proper. (Note that Dedekind's unique abstract structure (N, ϕ) of *the natural numbers* is not the same one whose existence he (almost) proves!) The appeal to a special predicate of concreta is merely a step in the justification of the general modal-existence postulate. Reasoning in the theory is not construed as reasoning "about the possibilities of marks" or anything so apparently irrelevant to pure mathematics. Furthermore, the predicate is used counterfactually. (Dedekind too is naturally read in this way: it is *in principle possible* to entertain in thought any thought that has already been reached, quite apart from how far anyone will actually ever get in this iterative process. We have preferred to appeal to a constructive rule for generating sequences of marks as a somewhat "purer" example, involving merely the "capacities" of a Turing machine (say, to add a stroke to the first blank cell encountered when moving along the tape in a given direction).) To complete the derivation, we have appealed to Fregean abstraction in the modal context of "Potential Infinity": there could exist strokes satisfying these conditions, and then there would be the class of those strokes, just as there would be the class of *individuals* satisfying any condition. Comprehension with respect to first-order objects is completely general, and it applies regardless of the size of the collection (or whole) in question. However, the fact that comprehension has been employed modally allows us to comply with Russell's requirement of "no contingent infinities" for mathematics. (The terminology "actual infinite" is, of course, confusing in this context. Comprehension tells us that there *could* be an infinite totality (of some sort that we need not settle here), and this is "actual" in the sense of "completed", as opposed to "potential". But, unfortunately, "completed" is even more confusing in its own way in suggesting unintended absurdities, e.g. that a mechanism designed to generate arbitrarily long finite sequences (in time and with bounded frequency) could at some time have generated an infinite sequence. Some have actually taken classical infinitistic mathematics to be committed to some such absurdity.)[23]

To conclude this section, it is appropriate to compare the modal-structuralist view with standard set-theoretic platonism on certain fundamental issues. Both accept as "meaningful" the highly non-constructive notion of "all subsets of an infinite set" in one form or another. (We have purposely left open the possibility of, for example, concrete interpretations of the ms second-order quantifiers as in the induction axiom, according to which "parts" rather than subsets of certain totalities are recognized.) Both recognize "second-order logical consequence" as having determinate sense, apart from our (in)capacities to verify or falsify particular cases. And, in rejecting deductivist accounts, both rest with certain irreducible "mathematical existence" postulates which seem to defy all efforts to sustain them as "analytic" or "true in virtue solely of linguistic meaning". Modal-structuralism seems no more able to guarantee the possibility of ω-sequences by the meanings of its words than Anselm could guarantee the existence of his deity by the meanings of his. Nevertheless, the modal-structuralist and platonist treatments of mathematical existence are quite different in interesting ways. The platonist can claim to eliminate problematic modal notions by means of straightforward (?) existence claims: the characteristic ms modal-existence theses are understood in terms of the existence of models (e.g. (1.9) is accepted because of prior acceptance of the existence of a standard model of PA^2, set-theoretically construed). However— setting to one side questions concerning the "univocality of 'exists'" on which this reduction relies, and also the familiar questions concerning reference and epistemic relations—there is a significant difference on the more directly mathematical question of the status of

---

[23] See again Dummett [1977], pp. 57–60.

infinite totalities brought out by the above line of argument. For the price the set-theoretical platonist pays for the elimination of modality is the need to postulate the axiom of infinity as a brute fact. (This is clear in the context of the Zermelo–Fraenkel system: the first infinite level behaves as a strongly inaccessible cardinal, i.e. all the other axioms hold in the structure of the finite ranks.) On the ms view formalized in modal second-order logic, the relevant sort of existence of infinite totalities is not an additional extra postulate, but, as indicated, follows from the two more basic assumptions, what we have called "Potential Infinity", and second-order comprehension, each of which has "constructive roots" in its own way. Potential Infinity, of course, has its roots in the most elementary procedures of "adding one", with which all constructive mathematics begins. (To what extent spatial or other geometric intuitions may be involved can be left moot.) And comprehension, of course, is tied to linguistic construction, i.e. the construction of predicates of the appropriate sort (although, of course, impredicative "construction" in this sense is allowed). Both these constructive principles ought to enter very early in the foundations of mathematics, even if they are to be transcended. The ms approach can claim to provide a way of accomplishing this which is actually fruitful at least to the extent already seen: combining the two constructive principles immediately yields the most fundamental infinite totalities required by classical analysis.

§ 3. Justifying the Translation Scheme

Turning back to the hypothetical component, we now confront the task of showing that the translation schemes (1.7) and (1.8) are accurate and adequate. What do we mean by this? We have already seen that the practice of theorem proving is respected. But, from a realist perspective, much more needs to be shown. What we should like to demonstrate is that, for any sentence A of the original arithmetic language ($\mathscr{L}(\mathrm{PA}^1)$ or $\mathscr{L}(\mathrm{PA}^2)$), A_p and A_{msi} "are fully equivalent for mathematical purposes", where "discovering truth" counts as a mathematical purpose. (Here, 'A_p' stands for A on a literal platonist reading.) Thus, "demonstrating accuracy" means that we should be able to show that in some suitable sense, A_{msi} "holds" iff A_p "holds", i.e. the translation preserves truth. And adequacy would simply mean that this pertains to all sentences of the original language.

But here we confront a serious problem. What is the standard of truth to be? On the platonist view, "truth" means "truth in the standard model" (either a unique model of "the natural numbers", or a fixed set-theoretical model), whereas on the modal-structuralist view, "truth" means—roughly—"truth in any possible model", where this is spelled out as truth of the relevant counterfactuals. Since the modalist does not offer a model-theoretic reduction of this latter notion—not having available the set-theoretic framework for carrying it out—the notion of truth as applied to the counterfactuals themselves is just disquotational. Moreover, the modalist recognizes no (actual) standard models at all; at least, a strict neutrality on the question is maintained. Thus, it should be compatible with the modalist position that all the platonist mathematical sentences are, strictly speaking, false. What is offered, in A_{msi}, is a replacement for A_p, not a genuine equivalent. Literally, how *could* the modalist accept "A_p iff A_{msi}"? Similarly, the platonist may reject the modal notions (as a Quinean platonist would); in that case, the equivalence would be equally out of reach.

The point is that the modalist and the platonist are operating in very different frameworks when it comes to evaluating a translation scheme as truth-preserving or not. In what framework is a demonstration of accuracy of the translation scheme to be carried out?

One might hope to define a "common core" system, a set of principles that both positions accept, and then to carry out a proof of equivalence within that framework. But here one may be asking for the impossible. There may simply be no system at all which (a) is capable of proving that A_p and A_{msi} are mathematically equivalent, and (b) consists entirely of assumptions acceptable to both viewpoints. (Such situations are almost typical in philosophy. Think of phenomenalist reduction programmes vs. ordinary (physicalist) material object discourse; or of classical logic vs. quantum logic, and so on.)

How are we to deal with this dialectical impasse? Rather than insist on a common core system—which may be asking for the impossible—, our strategy is to live with the impasse, and to let each side have its say separately. If the platonist can understand the modalist well enough to prove equivalence in the platonist framework, that would at least eliminate platonist objections to the msi on mathematical grounds. Moreover, that might carry interesting philosophical implications, for example, that the literalist semantics associated with platonism is unnecessary and misguided.

On the other hand, it would be complacent of the modalist to rest with such an outcome. If the view is to be philosophically satisfactory, it must at least be capable of its own "internal justification". One must at least be able to see, from within, that the truth-determinateness of the original sentences is respected. For if this could not be seen without putting back on the platonist's glasses (which do help one see so much!), the modalist position would have to be judged "dialectically unstable" in its own way. The upshot would be a kind of "resonating philosopher", forever flipping back and forth from one view to the other in the manner of the quantum field theoretician with his wave models (coupled harmonic oscillators) and particle models (constant creation and annihilation). If such dizziness can be avoided, the effort should be made.[24]

Let us first look at the matter from the external, platonist point of view. Looking back at the schemes (1.7) and (1.8), apart from the question of interpreting the modal operator, it should be fairly obvious that, from the platonist perspective, if an original arithmetic sentence, A, holds in the standard model (*sui generis* number-theoretic, or set-theoretic), A_{msi} should hold also, that is, the part following the box, call it A^-_{msi}, is just (the relativization of) either a truth or a falsehood of second-order logic. Either A holds in the standard model, N, or it doesn't. If it does, then, since all full models of PA^2 are isomorphic, A holds in all of them; if A fails in the standard model, then it fails in all full PA^2 models, i.e. not only is A^-_{msi} false, but $(\sim A)_{msi}$ is true. And, of course, the platonist agrees that "not both A^-_{msi} and $(\sim A)^-_{msi}$", since N is presupposed. Thus, using standard model-theoretic reasoning, the platonist sees that, apart from the modal operator, the translation schemes are fully bivalent and truth-preserving. (All of this reasoning, of course, could be formalized in set theory, either with the natural numbers as urelements or in pure set theory, taking ω together with ordinal succession as the

---

[24] The stance of the "resonating philosopher" is suggested by some remarks in Putnam [1967]. This may have been based on the belief that no single framework could really stand on its own, in that one would have to move outside any one in order to answer questions it prompted. Of course, no system can be universal in the sense of providing answers to all demands for justification. All systems have their starting-points. But we are concerned here with questions that an interpretative framework ought to be able to answer if it is to be convincing at all.

standard model.) All that is missing is a way of handling the modal operator.

But with the resources of set theory, the platonist confronts no insurmountable problem here. It is understood that a logico-mathematical modality is intended, supporting the S-5 axioms. It is further understood that, in the modal translates, all relevant conditions are explicitly set forth in the antecedents of the conditionals. That is, it is not necessary to have recourse to *ceteris paribus* clauses, or to any notion of relative similarity among possible worlds in evaluating these counterfactuals. They behave in accordance with the principles of strict implication, and, in this respect, are to be clearly distinguished from ordinary or causal counterfactuals. As is well known, the latter are highly sensitive to assumptions concerning "relevant background conditions", and this sensitivity is the source of deep problems in developing a semantics or theory of truth for these idioms.[25] However, any objections that a platonist may have to counterfactuals on these sorts of grounds simply do not carry over to the mathematical counterfactuals.

This suggests that the platonist may make reasonable sense of the modality in question by providing a set-theoretic semantics for it, without having recourse to extra machinery beyond set-membership. This would, in effect, provide a translation of the modal translates back into the language of set theory, and this could be used to compare A_{msi} with the original A.

In fact, the task of developing an appropriate semantics for logical modalities has already been carried out in sufficient detail for our purposes.[26] On this semantics, the part of worlds is played by model-theoretic structures of a given type, built on a given, fixed domain. Since the intended notion of logical possibility contemplates as rich a variety as possible of such structures, one is led to the "primary semantics" (in Cocchiarella's sense), in which all (set-theoretically possible) structures of the appropriate type, over the given domain, are assumed to be in the set of worlds of the model structure (for the modal language).[27] Such model structures are called "full" (not to be

[25] Recall Goodman's [1955] "problem of cotenability"; and reflect on the problems associated with a global similarity relation on possible worlds, designed to solve that problem (e.g. Stalnaker [1968], Lewis [1973]).

[26] See Cocchiarella [1975], [1984].

[27] Under the primary semantics, even modal propositional logic is incomplete in the sense that, provided the language contains a relational predicate, the set of logically true formulas is not recursively enumerable. (See Cocchiarella [1984].) But

confused with the "fullness" of the models of the second-order non-modal language (of PA), which has to do with whether the range of the second-order quantifiers (of degree k) is the full power set of the (k-fold Cartesian product of the) domain). The only adaptation we need make of Cocchiarella's primary semantics is to stipulate that the structures for our non-modal, mathematical language—the relevant worlds—are themselves to be full, with respect to the ranges of the higher-order quantifiers. As already indicated, of course, all worlds are mutually accessible.[28] Summing up, a modal sentence S in the second-order quantified modal language, $\mathscr{ML}(PA^2)$, based on $\mathscr{L}(PA^2)$, is valid or logically true just in case it holds under all assignments in all full free model structures based on a given infinite domain, in which the worlds of such model structures are full second-order structures, as already stipulated.

Now if we consider sentences A of $\mathscr{L}(PA^2)$, we have the following simple connection between A and its modal translate, A_{msi}, interpreted according to the semantics just sketched:

$$PA^2 \text{ logically implies A iff } A_{msi} \text{ is a (modal) logical truth.} \quad (1.10)$$

Logical implication on the left is just the usual model-theoretic notion with respect to full second-order (non-modal) logic, and logical truth on the right is the notion just introduced for modal sentences. The connection is just a matter of tracing definitions. (It is entirely unsurprising, since $\wedge PA^2$ is built into A_{msi}.)

Trivial though it may be, (1.10) does constitute a platonist *equivalence theorem*, in that, together with the well-known facts concerning second-order (non-modal) implication reviewed above, it shows the ms-translation scheme to be accurate and adequate. In this sense, the platonist can come to see that absolute reference to "the natural numbers" is superfluous. But equally important philosophically, and more interesting technically, is the question whether the modal-

axiomatic completeness is something the msi forswore long ago, when it recognized full second-order (non-modal) implication.

It should perhaps be mentioned that under this "primary semantics", certain unusual phenomena arise in the connections between syntax and semantics. For example, '$\Diamond p$' is counted as valid, but '$\Diamond(p \,\&\, {\sim}p)$' is counted as invalid, so that substitution for propositional variables does not preserve validity. Such things are anomalous, but I do not think they undermine the coherence of the semantics.

[28] Note, however, that different worlds of a model-structure can have different domains (whose union is the domain of the model-structure), and the Barcan formula can be avoided while satisfying the S-5 axioms, in the manner of Kripke [1963]. Such model-structures are called "free".

structuralist can arrive at a suitable equivalence theorem from within. Let us now consider this.

§4. Justification from within

The first real task is to formulate in a precise way just what the modal-structuralist should be able to prove with regard to the translation schemes (1.7) and (1.8). Obtaining (1.10) directly is out of the question, since it relies on general model-theoretic notions of truth and satisfaction (for the modal and original non-modal mathematical languages), and these are not available in their usual form in the restricted ms framework.

A brief comparison with Dedekind's procedure, in 'Was sind und was sollen die Zahlen', is again instructive. After developing his definition of "simply infinite system" (i.e. ω-sequence, i.e. satisfying the PA^2 axioms), Dedekind showed that any two simply infinite systems are isomorphic, i.e. between any two there exists a bijection preserving the successor relations. In modern treatments, this corresponds to the proof that the PA^2 axioms are *categorical*: any two full models of the PA^2 axioms (equivalently of the PA^{2+} axioms) are isomorphic. This, of course, is stated and proved in a set theory powerful enough to develop the formal semantics of $\mathscr{L}(\mathrm{PA}^2)$. Dedekind, of course, did not have that. Still, he gave a mathematically good proof of a good theorem:

> If X, f and X', f' are two arbitrary simply infinite systems, then there exists a map $\phi: X \to X'$, ϕ 1–1 and onto X', such that for any n, m in $X, f(n) = m$ if and only if $f'(\phi(n)) = \phi(m)$.   (1.11)

Now one way of formalizing this would be to have recourse to model theory: "X, f is a simply infinite system" would be rendered as "$X, f \vDash \wedge \mathrm{PA}^2$", etc. But this is not really necessary. Since only finitely many conditions are involved, one can simply write out the PA^2 axioms as conditions on X and f, i.e. one says directly that X has a (unique) "first element" (i.e. a z such that $X(z)$ & $x \neq f(y)$, any y such that $X(y)$), each element has a unique "successor" (written out similarly in terms of X and f), and so forth. But this is just what we have written as $(\wedge \mathrm{PA}^2)^X\!\left(\begin{smallmatrix} s \\ f \end{smallmatrix}\right)$, in the language of second-order logic. Thus, Dedekind's version of the *categoricity theorem* can be stated directly in second-order logic. (It is obvious that the consequent of (1.11) can be. Note, incidentally, that if we were working with PA^{2+}, similar clauses expressing that ϕ preserves addition and multiplication would be added. Nothing in the present discussion turns on which version we consider; for notational simplicity, we will work with PA^2.) Moreover, an analysis of the proof reveals that it can be carried out entirely within second-order logic (making use of instances of the comprehension scheme with parameters).[29] The result of these reflections is this: *The categoricity theorem (for PA^2) in the form (1.11) can be proved within the structuralist framework we have been considering throughout.* And this form is a direct and natural way of making precise what Dedekind originally proved.

But, one may ask, where is the element of modality? (1.11) appears to quantify over ω-sequences, and would be vacuous if in fact there aren't any, etc. However, there is a natural remedy. The modal-structuralist carries out the argument by beginning, "Suppose X, f and X', f' were any two ω-sequences; then there would be a mapping ϕ such that ..." Thus the result that is proved is not (1.11) *simpliciter* but the necessitation of (1.11). This is the most elementary modal logical step. Apart from it, modal logic plays no role in the proof. That, as I see it, is welcome news, since so much controversy surrounds the modal idioms. On the current approach, they are not entirely dispensable in mathematics, for they enable us to mark crucial distinctions and dispose of pseudo-questions (e.g. "What sort of object is 17?" etc.—recall our initial motivation). However, beyond this, the less heavily we rely on modalities in our mathematical reasoning, the better. So far, then, so good.[30]

---

[29] One obtains the existence of ϕ with the relevant properties in the Frege–Dedekind way, as "the intersection of all relations such that ...", where the blank is filled with the defining conditions on ϕ (sending the initial element of X to that of X' and "preserving successor"). This is all expressible in the language of second-order logic.

[30] Note that a modalist who takes seriously talk of possible worlds may be dissatisfied with the necessitation of (1.11), preferring instead a more complex modal formulation, such as,

$$\Box \forall X \forall f \Box \forall X' \forall f' \Box (\omega(X,f) \,\&\, \omega(X',f')) \supset \text{etc.}),$$

which is more general in considering the ω-sequences as occurring in different worlds. $\Box$ (1.11) in effect just looks at ω-sequences "within a world". But on the more general formulation, what is a relation or function between items in different worlds? These would be intensions, recognized as objects in their own right, and our logic should then admit modal formulas into the comprehension scheme, giving rise to such

Now one may be tempted to suppose that, in recovering the categoricity of PA^2, the structuralist has accomplished whatever could reasonably be demanded by way of an internal justification for the translation schemes. For (1.11) is a direct way of saying, within the second-order framework, that our axioms characterize a unique type of mathematical structure; obviously, then, it does not matter "which one" we are "talking about" when we are doing the mathematics of such structures. Isn't our justification complete?

It would be pleasant to conclude this, but overly sanguine. For, while the inference just drawn from (1.11) may indeed be intuitively obvious, really it demands a proof. For the inference pertains to language used to describe the structures, viz. the sentences of $\mathscr{L}(PA)$; yet (1.11) itself says nothing about these sentences. And, remember, it is a translation scheme—a representation of sentences of a given mathematical language—that is to be justified. There is thus a further step, from categoricity to a claim involving language, that needs to be taken.

Now, in a model-theoretic treatment, the step is well known. It is stated in the form of a general theorem, that isomorphic structures are elementarily equivalent, i.e. that they satisfy the same sentences (of the language on which the structures are based). That is, one proves,

$$M \approx M' \supset (M \vDash A \equiv M' \vDash A, \text{ all sentences A}), \qquad (1.12)$$

known as the *isomorphism theorem*. It is really the combination of the categoricity and isomorphism theorems that provides the full justification of the intuitive assertion that "it doesn't matter which ω-sequence we are talking about". And, indeed, if the set theorist argues rigorously that, for any A of $\mathscr{L}(PA^2)$, either A or $\sim$A is logically implied by the PA^2 axioms, both theorems will be employed. But this is precisely the argument that the platonist can give for the full bivalence of the translation scheme.[31]

---

monsters as "the class of all possible individuals", etc. If modal-structuralism were forced into this position, it would surely represent no gain at all. Our strategy must be to bypass all such modal complexities and attendant strong platonist commitments.

[31] It should be noted that Dedekind, although lacking Tarskian model theory, was sensitive to the problem of justifying the "transfer of language" from one ω-sequence to another. For he followed his categoricity theorem with a kind of converse:

X, f an ω-sequence & $X, f \approx Y, g \supset Y, g$ is also an ω-sequence

Now (1.12) involves a satisfaction relation not available to our modal-structuralist. Still, we may ask for the next best thing, namely a proof, for each sentence A of $\mathscr{L}(\text{PA})$ (at least first order, better second order), of

$$A_{\text{msi}} \vee (\sim A)_{\text{msi}} \text{ (and "not both")}, \tag{1.13}$$

which is a direct expression of bivalence (as the connectives are understood classically). (We have already seen that the parenthetical "not both" presents no problem. Here we focus on the disjunction.) This is a natural and reasonable substitute for (1.12); while this generalization over sentences is beyond the framework, (1.13) is firmly within it, and approximates (1.12) in the way in which a proof of each instance of a theorem scheme approximates a metatheorem which says that every instance is provable (or is true).

Now, in fact, there is a way in which (all instances of) (1.13) can be obtained. It is naturally broken down into two steps. First one proves the following:

Elementary Equivalence Theorem: In standard axiom systems of second-order logic, the following is derivable:

$$\forall X \forall f \ldots \forall X' \forall f' \ldots \left\{ \omega(X,f \ldots) \,\&\, \omega(X',f' \ldots) \supset \right.$$

$$\left. \left[A^X\!\begin{pmatrix} s \ldots \\ f \ldots \end{pmatrix} \equiv A^{X'}\!\begin{pmatrix} s \ldots \\ f' \ldots \end{pmatrix} \right] \right\} \tag{1.14}$$

for each sentence A of $\mathscr{L}(\text{PA}^1)$, in which case, fill the '...' with 'g' and 'h', and construe '$\omega(X,f,g,h)$' as '$(\wedge \text{PA}^{2+})^X\!\begin{pmatrix} s, \Sigma, \Pi \\ f, \ g, \ h \end{pmatrix}$', or, in the second-order case, of $\mathscr{L}(\text{PA}^2)$, in which case, drop the '...', and read '$\omega(X,f)$' as '$(\wedge \text{PA}^2)^X\!\begin{pmatrix} s \\ f \end{pmatrix}$'.

This is a non-metalinguistic way of saying that the sentence A is treated in the same way by any two ω-sequences. (It is our approximation to the model-theoretic claim of elementary equivalence.) The proof of this is somewhat laborious, but follows as the limiting

---

(see Dedekind [1901], § 133), i.e. that any structure isomorphic to one satisfying the PA^2 axioms also satisfies the PA^2 axioms. This is a step on the way to proving elementary equivalence of isomorphic structures, but of course it is not the completed result.

case of a corresponding lemma involving formulas with free variables, proved by induction on the complexity of A, utilizing the categoricity theorem (1.11).[32]

Once we have (1.14), hence its necessitation (call it (1.14′)), there is still the task of getting to (1.13). In this step, we must attend to the behaviour of the modal operators, and, in fact, we need a further (mild) assumption governing their behaviour.

To see the problem, proceed by *reductio*, assuming $\sim[A_{msi} \vee (\sim A)_{msi}]$. This gives us something of the form,

$$\Diamond \exists X \exists f \dots (\omega(X, f \dots) \& \sim B)$$
$$\& \Diamond \exists X' \exists f' \dots (\omega(X', f' \dots) \& B') \tag{1.15}$$

(where we let 'B' abbreviate '$A^X\!\begin{pmatrix} s \dots \\ f \dots \end{pmatrix}$', and similarly for 'B''). But to contradict (1.14), we need to arrive at

$$\Diamond \exists X \exists f \dots \exists X' \exists f' \dots (\omega(X, f \dots)$$
$$\& \, \omega(X', f' \dots) \& \sim[B \equiv B']). \tag{1.16}$$

---

[32] In the first-order case, the lemma is as follows:

Let 'Hyp' be the conjunction of $\omega(X, f, g, h)$, $\omega(X', f', g', h')$, and the statement "ϕ is a bijection from X onto X' & ϕ preserves f, g and h"; then, for each formula $A(x_1 \dots x_k)$ of $\mathscr{L}(PA^1)$ (all free variables displayed), the following is derivable in axiomatic second-order logic plus Hyp:

$$\forall x_1 \dots \forall x_k \left\{ X(x_1) \& \dots \& X(x_k) \supset \left[A^X(x_1 \dots x_k) \begin{pmatrix} s, \Sigma, \Pi \\ f, g, h \end{pmatrix} \right] \equiv \left[A^{X'}(\phi(x_1) \dots \phi(x_k)) \begin{pmatrix} s, \Sigma, \Pi \\ f', g', h' \end{pmatrix} \right] \right\}. \quad (*)$$

Where A is a sentence, this reduces to the consequent of (1.14). The second-order case is similar, employing the appropriate formulation of relativization to X in the antecedent of (*) (see n. 18) and reference to "the relation induced by ϕ on the relation R_i" in the consequent.

Note that these lemmas allow a proof of each instance of (1.14) within the ms framework, which is weaker than the general statement of the theorem itself, viz. that each instance is provable. To obtain this as well, it would be necessary to provide a ms treatment of syntax, either by means of Gödel numbering within any hypothetical ω-sequence (a fixed domain of "the natural numbers" must be avoided), or by means of a direct theory of syntax, say along the lines of Belnap [unpublished], utilizing suitable structural induction principles, and modalized so as to avoid any categorical commitment to (infinitely many) syntactic objects. (Talk of such is replaced by talk of what would hold of any objects that satisfied the principles of syntax.) While it would be tedious to write all of this out, I see no obstacle in principle to its being done. In these ways, then, the structuralist can assert and derive the general (proof-theoretic) result, and is not confined to the instances.

This would follow from (1.15) by prenexing and sentential logic if the modal operators were absent. But with them, the move is not in general valid. We submit, however, that in the pure mathematical cases of interest, the move is harmless. In fact, it is a special case of a general principle which we already have implicitly accepted in formulating categoricity as we have ((1.11) above). Let A and B be any two sentences in the original arithmetical language, i.e. either sentences of $\mathscr{L}(PA^1)$ or of $\mathscr{L}(PA^2)$. The principle asserts that if A can hold in an ω-sequence, X, f, and if B can hold in an ω-sequence, X', f', then it is possible that A holds in X, f and that B holds in X', f', i.e.

$$\Diamond\left[\exists X \exists f \ldots \left(\omega(X,f\ldots) \ \& \ A^X\!\left(\begin{matrix}s\ldots\\f\ldots\end{matrix}\right)\right)\right] \ \&$$

$$\Diamond\left[\exists X' \exists f' \ldots \left(\omega(X',f'\ldots) \ \& \ B^{X'}\!\left(\begin{matrix}s\ldots\\f'\ldots\end{matrix}\right)\right)\right] \supset$$

$$\Diamond\left[\exists X \exists f \ldots \left(\omega(X,f\ldots) \ \& \ A^X\!\left(\begin{matrix}s\ldots\\f\ldots\end{matrix}\right)\right)\right.$$

$$\left. \& \ \exists X' \exists f' \ldots \left(\omega(X',f'\ldots) \ \& \ B^{X'}\!\left(\begin{matrix}s\ldots\\f'\ldots\end{matrix}\right)\right)\right], \qquad \text{(AP)}$$

which may be called an accumulation principle. To use the possible worlds metaphor, any two possible ω-sequences may be accumulated in a single world. (Of course, the primes on the variables play no role, but they would be introduced in moving to prenex normal forms, so we have put them in in advance.) It was just this idea that led us to rest with (1.11): if any two possible mathematical structures may be thought of as "in a single world", then we may speak of relations or mappings between them in the usual way, without further assumptions regarding intensions. The scheme (AP) gets at this directly, without ascending to talk of worlds. Its justification lies in the fact that the mathematical sentences (A, B, etc.) concern only what is internal to an ω-sequence; in a rather precise sense, they give rise only to "internal relations", as expressed in their relativizations to single ω-sequences in (AP). Thus, no conflict between two such possibilities can arise—as it could if we had to deal with non-mathematical properties such as "being the only ω-sequence created by Dedekind"—and the conjunctions in question are then also possible. (Note the effect of quantifier relativization: in the second-order case, attempted reference to "all ω-sequences" comes out as "all ω-subsequences of X".)

But now we are home. For now an appropriate instance of AP will take us from (1.15) to (1.16), contradicting (1.14) outright. Thus, from any instance of (1.14), the corresponding instance of (1.13) can be derived, completing this much of the internal justification procedure.

Within the ms framework, it can be seen that the translation scheme is fully bivalent. But can it be seen to be accurate in the sense of "getting the right answers", as the platonist understands this? Yes, in a sense, this too can be seen, by a very elementary argument. It is simply that the ms translate, A_{msi}, of A already expresses an accurate standard of truth from the platonist point of view. Although the platonist might articulate a *preferred* standard in terms of a unique fixed ω-sequence ("the standard model N")—something which the modal-structuralist wishes to avoid—, the platonist will also regard it as accurate to say that

A is platonistically true iff PA^2 logically implies A,

in the sense of full second-order implication. But, if the platonist is prepared to reason modally at all, he will also accept,

PA^2 logically implies A iff A_{msi}.

Thus,

A is platonisitically true iff A_{msi},

is something the platonist can accept. Thus, the modal-structuralist, as well as the platonist, can see, by means of a standard acceptable to the platonist, that platonist truth is respected by the translation scheme. In sum, the modalist can see from within that the proposed representation is completely faithful for mathematical purposes.

§5. Extensions

Turning now, briefly, to more powerful theories, it should be clear that the above analysis can be carried out in parallel fashion for real analysis, RA^2, in which first-order variables (as usually interpreted) range over real numbers and second-order variables over sets (and relations) of these. As is well known, in order to obtain a categorical system, the continuity principle, or least-upper-bound axiom, must be stated in second-order form:

$$\forall S(S \text{ a non-empty bounded set of reals} \supset S \text{ has a l.u.b.}). \qquad \text{(C)}$$

Beyond this, there are different axiom sets that lead to the same result. One could take the theory to be continuity (C) together with the usual first-order field axioms (F) and order axioms (O). Alternatively, one could take the theory to be (C) together with axioms (O′) for a dense linear ordering without end-points, and the axiom of separability (S), stating that there is a countable dense subset. In the latter case, "countable" can be spelled out by the statement that there exists an ω-sequence together with a bijection between it and the dense subset. This can all be stated in second-order logic. In either case, one can recover the "categoricity" as a pure, second-order mathematical theorem, analogous to (1.11). (It is an exercise to check that the usual arguments, including the Cantor back-and-forth proof that the structures in question contain "copies of the rationals", do not transcend second order.) Furthermore, the internal argument of the last section for accuracy of the translation scheme carries over intact (for both first- and second-order sentences).

Of course, the argument for bivalence employs the relevant modal existence assumption,

$$\Diamond \exists X \exists f (\wedge \text{RA}^2)^X \binom{<}{f}, \qquad (1.17)$$

asserting the possibility of (say) complete ordered separable continua. Like modal existence for PA^2, (1.9), this is a strong assumption not reducible to a claim of formal consistency. But, unlike (1.9), (1.17) cannot be derived from anything with a claim to constructivity such as was seen in "Potential Infinity". No doubt (1.17) has its roots in our geometric experience, but the idealization motivated by such experience is far greater in the case of (1.17) than in that of (1.9). Even more than (1.9), (1.17) must be regarded as a working hypothesis of classical mathematics, not as a self-evident certainty.

It should be noted that RA^2 is, in effect, third-order number theory, and involves quantification over arbitrary subsets of reals. It is a very powerful theory, indeed. By making use of coding devices, virtually all the mathematics commonly encountered in current physical theories can be carried out within it.[33]

As indicated at the outset, there is a significant tension between structuralism and logicism, highlighted by their diverse treatments of analysis. Logicism, of course, provides a reduction to set theory (or type theory): all the primitives of RA are defined and all the axioms are derived. Structuralism, on the other hand, reflects the viewpoint of many texts which simply present an adequate list of axioms and say that "real numbers are whatever satisfy these axioms". It is worth pointing out, however, that, to some extent, the structuralist can respect the logicist analyses. This emerges especially clearly in connection with the choice of systems just sketched. If RA^2 is taken in the second way (as the theory of separable ordered continua), it will be necessary to introduce the field operations and recover the field axioms as theorems. This is naturally accomplished along logicist lines. To summarize the construction: One is given an ω-sequence, (X, f) (by the hypothesis of separability), and, via a number-theoretic coding, one can talk of pairs, triples, etc., of natural numbers. This enables introduction of the rationals Q_X (relative to X) and the algebraic operations among these. By the Cantor back-and-forth argument, Q_X can be put in one–one order-preserving correspondence with the (given) dense subset of the domain of "reals", and this can be extended to a correspondence between Dedekind cuts in Q_X and those reals. Finally one uses this correspondence to transfer the logicist algebraic operations on the cuts to the reals in question. (One can then go on to define an ordering on the reals in terms of the algebraic operation of addition and then show that it agrees with the original assumed ordering.) Thus, a good portion of the logicist construction is respected and utilized in the structuralist account.[34] The main item sacrificed, of course, is the derivation of continuity (C)— the structuralist must simply assume it as part of the characterization of continua. On the other hand, set-theoretic assumptions have been kept to a minimum. In fact, except for the classes and relations

---

[33] Cf. Burgess [1984].

[34] It should be noted that the recovery of ordinary use of individual constants for rationals or irrationals depends on some construction or "means of introduction" such as the one just outlined. In order to "fix the reference" of, say, '$\sqrt{2}$', one may appeal to a hypothetical ω-sequence together with a mapping between cuts in rationals constructed over it and the individuals of any hypothetical separable ordered continuum. Naming is thus relative to a construction, and there need be no absolute "positions" (cf. above, Introduction, n. 12).

among individuals assumed in the second-order comprehension axioms, the structuralist ontology as we have presented it need not transcend that of a strict nominalism. We conclude with a few remarks on the relation between modal-structuralism and nominalism.

§6. The Question of Nominalism

As we have seen, modal-structuralism abstracts entirely from the "first-order objects"—i.e. objects in the range of first-order quantifiers—of mathematical theories such as PA and RA. In short, it dispenses with such objects entirely. All first-order content is captured in the structural relations in which any objects whatever would have to stand in order collectively to constitute a mathematical totality of the appropriate sort. As a result, ontological questions such as "What sort of entities are numbers?" fall by the wayside.

The same can hardly be said, however, with respect to the second-order content of mathematical theories. Both inside and outside of modal contexts, the ms interpretation quantifies over classes of, and relations among, individuals. Although it manages to avoid collections of these, and further collections of such collections, etc., still it countenances one "abstract level". And, one may well suspect, this is indispensable. For how could mathematical structures possibly be characterized without any talk of classes or relations?

Contrary to this suspicion, however, there is in fact a way in which even the single abstract level of second-order logic can be dispensed with. In effect, it is possible to read the ms interpretations given so far *entirely* nominalistically. This is worth examining for two reasons: first, it helps emphasize the vast difference between nominalism and constructivism, something that deserves to be better appreciated; and second, it is clearly relevant to "indispensability arguments" for abstract objects such as classes and relations.

The main idea is to restrict the consideration of hypothetical structures to those with which nominalism can deal. The nominalism at issue is a view which recognizes arbitrary sums of any individuals independently recognized.[35] This is partly formalized in the calculus

---

[35] For a nominalization programme that employs this "complete logic of nominalistic sums", see Field [1980]. In Field [1984], modality is explicitly introduced into the programme. The nominalization strategy outlined here is far more direct than that of Field [1980], and it respects the truth of (pure) classical analysis, in sharp contrast to

of individuals, in which the basic notion is a part–whole relation, $<$, or a relation of "overlaps", o, interdefinable with part–whole (via $x \, o \, y \equiv \exists z(z < x \, \& \, z < y)$, and $x < y \equiv \forall z(z \, o \, x \supset z \, o \, y)$).[36] In place of the usual second-order comprehension scheme, one has instead,

$$\exists x \phi(x) \supset \exists x \forall y(y \, o \, x \equiv \exists z(\phi(z) \, \& \, z \, o \, y)), \tag{CΣ}$$

the antecedent serving to avoid any null individual. (The notion of (non-empty) product of individuals is readily definable, and instances of the corresponding scheme, (CΠ), for products of all individuals satisfying ϕ are derivable.)

Now consider the case of arithmetic. Suppose we have a progression of concreta, strokes, stars, anything you like. Imagine them arranged in a linear order in space. The first-order quantifiers of the PA^2 axioms range over these concrete individuals. And the second-order (monadic) quantifier ranges over sums of them—arbitrary sums of them. Clearly, these are in one–one correspondence with the subsets. (This does not make the sums the same as the subsets, any more than the correspondence between individuals and their singletons makes these the same.) What guarantees this correspondence? The fact that each sum is uniquely decomposable into items forming the progression. But this was guaranteed by the fact that the initial items chosen were pairwise discrete (containing no common part). If overlapping were permitted, it could happen that distinct lists of initially given individuals formed the same sum. To ensure that this doesn't happen, one can add to the calculus of individuals an axiom of atomicity,

$$\forall x \exists y(A(y) \, \& \, y < x), \tag{AA}$$

where '$A(y)$', "y is an atom", is defined by,

$$A(y) \equiv \forall z(z < y \supset z = y).$$

Field's instrumentalism. The present approach can be extended to applied mathematics (which is the focus of Field's programme), via the idea of "embedding (portions of) the material world in a mathematical structure". This will be taken up in Ch. 3.

[36] For an exposition of the calculus of individuals, see e.g. Goodman [1977]. It should be mentioned that the calculus of individuals may not be the only framework within which to realize nominalist interpretations. The "plural quantification" conception of monadic second-order logic of Boolos [1985] may provide an alternative. There would be a gain in that discreteness or atomicity assumptions (needed on the present approach) could be dropped. Such an alternative merits further investigation.

It follows that all atoms are pairwise discrete. Any progression of atoms exhibits "enough" sums, i.e. distinct ones corresponding to each subset. Note, moveover, that these sums are, ontologically, on a par with the atoms themselves: they are every bit as concrete, and they are "as objective", i.e. independent of our powers of selection or "sum formation". They are simply parts of the whole progression, whether we select them—or like them—or not.

We have been speaking, intuitively, of concrete progressions, but how precisely is this to be expressed? We cannot yet simply write out the PA^2 axioms with the quantifiers interpreted as just mentioned, for what is to count as the successor relation? Monadic second-order quantification can be interpreted nominalistically, but here we have polyadic. How are relations in general to be handled within nominalist confines?

There are a number of approaches. One approach is not to quantify over two-place relations at all but simply to *use* certain predicates, in writing out the PA^2 axioms, which specify our intended meaning. We could, for example, use "is adjacent to", or "comes after", etc., relying on our ordinary understanding of such terms to specify the relevant type of structure. It may be complained that to proceed in this way is to make arithmetic and analysis dependent on geometric ideas, and perhaps the complaint is just. If indeed we are thinking of our mathematical structures as embedded in a geometric space, why not make this explicit in the well-known ways, building up number theory and analysis directly in terms of geometric primitives?[37] This is one familiar "nominalization strategy", although as usually presented it involves commitment to entities such as "points", which to many seem as problematic as numbers or sets themselves. In a modal-structuralist presentation, there is, of course, no commitment to actual points or other geometric objects. Instead, there is the modal-existence postulate, that spaces of the appropriate type are possible. The stuff of which they are composed is immaterial. (I acquiesce in the unintended pun.)

If one wishes to free number theory and analysis from geometry, one can, without transcending nominalism (i.e. modal nominalism). One may consider concrete, atomic structures (i.e. wholes) with further concrete atoms to serve as ordered pairs, triples, etc., of the

---

[37] See e.g. Blumenthal [1961]; also, Burgess [1984] and Resnik [1985].

original atoms.[38] In effect, one is considering product spaces, X^2, X^3, etc., based on the ground space X, but, instead of constructing them as one does in set theory (taking $\langle x, y \rangle$ as, say, $\{x, \{x, y\}\}$), one simply entertains additional individual atoms appropriately related to the original atoms. One is here adopting a "structuralist" approach to ordered pairs, implicit in mathematics texts which work with a notation for ordered pairs (say, "$\langle u, v \rangle$") without ever specifying any particular objects to serve as realization of the notation.[39] As everyone recognizes, all that matters is that the crucial law:

$$\langle u, v \rangle = \langle x, y \rangle \equiv u = x \,\&\, v = y,$$

be obeyed, together with the implicit existence assumption, that for any objects u, v, of the original domain of interest, there be something serving as ordered pair $\langle u, v \rangle$. Any way of correlating u and v with an object $\langle u, v \rangle$ obeying these laws is sufficient for mathematical purposes. For the purposes of nominalistic interpretations, this could be formalized by taking a three-place relation, $OP(x, y, z)$, as primitive (understood as "z is an atom correlated with x and y as their ordered pair"), and rewriting the uniqueness and existence requirements as axioms governing OP. The uniqueness requirement then takes the form,

$$\forall x \forall y \forall z \forall u \forall v \forall w [OP(x, y, z)$$

$$\&\ OP(u, v, w) \supset (z = w \equiv x = u \,\&\, y = v)];$$

and if one requires ordered pairs only for (say) atoms satisfying a condition X, the existence requirement is simply that

$$\forall x \forall y [X(x) \,\&\, X(y) \supset \exists! z (OP(x, y, z))].$$

Any interpretation of these axioms in terms of atomic individuals may be called a "concrete Cartesian space (of order 2) over X", and we may use standard mathematical notation ('$X \times X$' or 'X^2', or, more fully, '$\langle X, X^2 \rangle$' if we wish, reserving 'X^2' for "concrete Cartesian product") to refer to such a space, with the understanding that it is a concrete realization of the structural axioms that is intended, not the standard set-theoretic interpretation. The geometric interpre-

---

[38] This is, in effect, what Putnam [1967] did when he considered concrete models of Zermelo set theory; the relation of membership was given by further concreta— "arrows"—connecting the "points" of the model.

[39] See e.g. Dieudonné [1969], Ch. 1, § 3.

tations alluded to above would be specific examples of such spaces. Obviously, further axioms could be written down for concrete Cartesian spaces of higher order.

Given such a concrete Cartesian space, relations can be taken, nominalistically, simply as sums of tuples of the appropriate degree. Since the ordered tuples are stipulated to be atoms, such sums are in one–one correspondence with the subsets (speaking platonistically). In this sense, the nominalist is quantifying over arbitrary relations on a domain. If we are representing just PA^2, we only require X^2, for any domain X, in order to characterize the structures (since the successor relation suffices). But to represent arbitrary arithmetic operations, X^3 will be needed. This also suffices, for number-theoretic codings (definable from successor, plus, and times) can then be used to obtain ordered k-tuples, arbitrary k. For RA^2, just to characterize the structures, one requires just ordered pairs, in three roles: for the basic ordering on the ordered continuum; for the successor relation on an associated ω-sequence (assumed in the separability axiom); and for a mapping between the latter and a dense subset of the former. (And, of course, for set theory itself, one again requires only ordered pairs.) Thus, an enormous amount of mathematics can be represented in terms of concrete atomic structures with concrete Cartesian spaces of just these limited orders. And, of course, it is never asserted that spaces of these sorts actually occur. It is enough if they are logical possibilities. Then one may understand the mathematics in question as saying what must hold in any that there might be.

Finally, it should be pointed out that the model theory of these structures can also be carried out without positing more than Cartesian triple products. Consider number theory. As just indicated, X^3 suffices for interpreting arbitrary relations via coding (where X satisfies the PA^2 axioms). Further, the syntax of $\mathscr{L}(PA)$ can be coded in X by Gödelian techniques. But now arbitrary functions from syntactic variables (of first and second order) to objects of the appropriate sort are available, simply as sums of the right sorts of atoms of X. (Individual variables, x, are assigned atoms, a, of X; k-ary relation variables p^k, are assigned various k-tuples, u, of atoms of X.) An evaluation of (first- and second-order) variables will then be a sum of atoms of the form $\langle `x`, a \rangle$ and $\langle `p^k`, u \rangle$, (where quotation indicates "code of"). Thus, all the machinery for introducing satisfaction (as additional theory) is available within this framework. Similar remarks apply in the case of RA, since, once the algebraic operations on reals have been introduced, coding of k-tuples of reals become available; evaluations of (first- and second-order) variables can then be identified with nominalistic sums of reals, and the theory of satisfaction for these structures can be developed.

2
Set Theory

§0. Introduction

We have seen how a framework of second-order modal logic can serve to represent a structuralist interpretation of basic mathematical theories such as number theory and real analysis. To recapitulate the main points: on this interpretation, ordinary mathematical statements are construed as elliptical for hypothetical statements as to what would hold in any structure of the appropriate type, this being describable directly in second-order logical notation (using a sufficient, finite set of axioms, suitably relativized, and generalizing on the relational primitives of the theory). Absolute reference to mathematical objects is eliminated entirely. Instead, there is, in addition to the translation scheme (the "hypothetical component"), a categorical component to the effect that structures of the appropriate type are logically possible. This was found to be an irreducible, non-analytic working hypothesis associated with the mathematical practice codified in the theory in question (PA^2, RA^2, etc.). It was then shown how both the traditional "objects platonist" and the modal-structuralist—working from within their respective frameworks—could justify the interpretation as accurate and adequate. In particular, the modal-structuralist was able to establish the truth-determinateness of the translation scheme by recovering suitable versions of the categoricity of the theories in question, and the "isomorphism theorem" (that isomorphic structures satisfy the same sentences).

Here we confront the task of extending this sort of interpretation to set theory itself. There are a number of reasons for attempting this. First of all, it is of interest to know how far the approach can be extended. Does the msi constitute an alternative point of view in general, or is it merely an alternative with respect to theories of the sort already treated? Secondly, set theory represents both a great opportunity and a challenge to the approach; an opportunity since, as is well known, so much mathematics can be represented within set theory. In so far as set theory yields to a ms treatment, so does all

set-theoretically representable mathematics. (Thus, model theory—
of special interest to logicians, but not directly representable in the
second-order framework of the msi—would become available, at
least indirectly.) And set theory represents a challenge since, prima
facie, much of our ordinary talk and theorizing concerning sets
seems quite different from that concerning the number systems. One
tends to think of sets as absolute objects, independent of structure, in
marked contrast to numbers, which we quite readily conceive as
merely "positions in a structure" (in Resnik's [1981] terminology).
How are these intuitions to be accommodated?

Thirdly, there are certain puzzles—if not paradoxes—associated
with our standard platonist picture of a fixed, actual set-theoretic
universe, "*the* cumulative hierarchy". These come under the heading
of "proper classes", totalities "too big" to be sets—but not "too big"
to be very much like sets. For example, they contain members, i.e. are
collections of a sort; and they can even bear a (the?) membership
relation to "superclasses" (or "hyperclasses") obeying laws very
much like those of set theory.[1] Such systems are provably consistent
relative to certain natural strengthenings of ZF (e.g. ZF + "There
exists an inaccessible cardinal").

One can put the central problem raised by proper classes this way:
either we allow set-like operations (e.g. forming singletons, taking
power sets, etc.) to be performed on them or we do not. If we don't,
this seems an arbitrary restriction on those operations; and, in fact,
as just noted, we *can* (relatively) consistently extend such operations
to these "objects". But if we do "extend the operations", we find our
proper classes behaving just as a further level—an inaccessible level,
V_κ—of our initial cumulative hierarchy.[2] Initially, proper classes
were introduced (by Cantor, as "inconsistent totalities") in an effort
to capture a notion of potential infinity associated with ordinals. In
some sense, there are "too many" ordinals for their totality to form a
set. But—despite the formal convenience of proper classes and the
formal (relative) consistency of systems recognizing them—the
dilemma they present is troubling.

But wait, you may say: what's wrong with ZF, which simply dis-

---

[1] For examples of axiom systems formalizing such "continuations of ZF", see
Fraenkel, Bar-Hillel, and Lévy [1973], pp. 142 ff. In one case, the axioms for the
"superclasses" are exactly the ZF axioms!
[2] For a related discussion of proper classes in historical perspective, see Maddy
[1983].

penses with proper classes, and which is, after all, the central mathematical theory employed by set theorists? In response: as a mathematical theory, nothing is wrong with it; and, as we see it, it is perfectly right in refusing to countenance proper classes. However, the standard platonist picture associated with the theory is at odds with this refusal, for it takes the theory to be about a unique existing totality, "Cantor's universe". Perhaps one can steadfastly refuse to talk this way while still adhering to the standard picture, but it seems to be very difficult to do so. Consider the predicate "is a set", or "is an ordinal". In our overall semantics, we naturally wish to assign an extension to such predicates. But, on the standard platonist picture, such extensions would be proper classes. (Of course, they cannot be consistently treated as "sets" in the technical sense; but they would be recognized as totalities of *some* sort, and this is enough to generate the predicament just described.) It is worth attempting to develop an alternative picture.

In connection with proper classes, it should also be mentioned that there is a strong temptation to take them seriously for purely mathematical reasons. This shows up especially in connection with strong axioms of infinity and large cardinals. One important motivation for certain of these has been via reflection principles, which take the form: "The universe V has largeness property P; therefore there should be a level V_α of the cumulative hierarchy which also has P."[3] Any alternative view which rejects proper classes automatically must forgo this motivating strategy. How significant a sacrifice this is must then be considered. We shall return to this issue below (§ 5).

The core idea of an alternative view rejecting proper classes can be found in Zermelo's classic paper of 1930.[4] There, after setting forth what is essentially ZF set theory (but using second-order versions of Replacement and *Aussonderung*), and after proving a number of "quasi-categoricity theorems" (about which more below) intimately related to inaccessible cardinals, he concluded with some intriguing remarks on how the results just set forth could be seen as resolving the paradoxes associated with "inconsistent totalities". Set theory should be seen, not as the theory of a unique, all-embracing

---

[3] See Reinhardt [1974a]; also Kanamori and Magidor [1978].

[4] While the mathematical content of this important paper, Zermelo [1930], is apparently well known to set theorists, its philosophical content has not been sufficiently well appreciated, in part because, at the time of this writing, it still does not exist—published—in English translation. (It *has* been translated by Burgess [unpublished].) For a presentation of some of the main points, see Moore [1980].

structure, but instead as a theory of an endless infinity of intimately related structures. (These would be described in ZF as of the form $(V_\kappa, V_{\kappa+1} \in | V_{\kappa+1})$, for inaccessible κ; they are linearly ordered by end-extension; see below, §2.) In this way, Zermelo hoped to sustain two closely related informal principles; (i) the *generality* of set-theoretic concepts, summed up in the thesis that any (set-like) structure can be treated as a set; and (ii) the principle of *extendability*, that any set-theoretical universe can be viewed as part of a more comprehensive one.[5]

However, Zermelo's concluding remarks were only suggestive, and they have never been developed in detail. Moreover, it is, prima facie, not clear how they can be, since they embody a certain tension that can readily lead to contradiction. On the one hand, we have the principles just mentioned. But on the other, we have the apparatus of second-order logic—and, as is well known, it is essential to Zermelo's quasi-categoricity theorems that the theory (ZF^2) be formulated in second order.[6] But the second-order comprehension principle immediately yields proper classes, e.g. the class of all sets, or of all ordinals, or of all V_α, etc. Somehow, comprehension must be restricted.

A suggestion as to how this might be accomplished can be found in Putnam's controversial [1967]. Here, like Zermelo, Putnam regarded set theory as concerned with a multiplicity of structures, and he even (independently) formulated the extendability principle (in practically identical language, modulo translation into German!). However, Putnam's focus was on "nominalist" interpretations, in which modality would play a major role in enabling one to posit the possibility of sufficiently rich concrete structures (i.e. structures describable entirely within the calculus of individuals together with "nominalistically acceptable" predicates). But, apart from the question of nominalism,[7] modality can perform other functions. In par-

---

[5] An independent statement of this very principle was given by Putnam in his [1967].

[6] See below, §2. This is yet another illustration of the greater expressive power of second-order axiom systems. Cf. Shapiro [1985] and above, Ch. 1, §1.

[7] In his [1967], Putnam employed modality as the sole "non-nominalist primitive" in developing a view of "mathematics as modal logic", intended to be mathematically equivalent to the "mathematics as set theory" view. Thus, in his treatment of set theory, Putnam focused on "concrete models" (of Z rather than ZF, as it happened). As suggested in Ch. 1, this can be seen as a particular reading of second-order formulations, and, in order to leave open the possibility of such nominalist interpretations, we do not wish to transcend second-order logic. Nominalism, however, is not our focus here.

ticular, it can be used to make explicit our informal talk of "potential infinity" in connection with ordinals. When this is done in a suitable fashion, as we shall see below, a way of resolving Zermelo's dilemma emerges. This provides us, then, with a fourth motivating reason for attempting to extend modal-structuralism to set theory: to try to use modality to capture the informal distinction between actual and potential infinity in set theory, and to sustain—if possible— Zermelo's principles: generality, extendability, and the use of second-order logic.

§ 1. Informal Principles: Many vs. One

To guide our inquiry, let us list here, with explanatory discussion, what we take to be the main informal principles of modal-structuralist set theory. These are inspired primarily by the views of Zermelo and Putnam just cited, and represent our attempt at a synthesis, preparatory to a more formal treatment.

1. *Multiplicity and mathematical existence of structures:* Set theory, as embodied in ZF and related systems, is the study of structures of a certain type together with their interrelations. It is not to be conceived as the study of a single fixed universe. Of course, it is assumed that structures of the relevant type are logical possibilities. The relevant type is to be specified by means of axioms, and—in analogy with the ms treatment of arithmetic and analysis—such specification is seen as one of the primary functions of axiomatics.

Remarks: This, of course, embodies the Zermelo–Putnam rejection of the unique cumulative hierarchy in favour of a multiplicity of related structures. As we shall see, it is natural, from a classical point of view, to take these to be (isomorphic to) the so-called "natural models" of set theory (see below, § 2). (This turns out to be the most direct generalization (of which we are aware) of the treatment of PA and RA.) It also embodies a fundamental assumption of mathematical existence, stated modally as in the cases of PA and RA. Such a categorical assumption we take to be implicit in classical mathematical practice. It appears to be irreducible to any sort of linguistic convention, and serves to distinguish modal-structuralism from the varieties of "if-thenism". (For further discussion of such principles, cf. Chapter 1, § 2.)

2. *Non-absoluteness of "sets as objects"*: Sets are treated as such in

virtue of their role in a structure of the relevant type. They need not be thought of as having an absolute identity. In particular, ordinals are not taken as absolutely identified, but are treated only as ordinals of a model.

Remarks: It is a commonplace to regard the identification of ordinals with particular sets as merely a convenience. They serve to represent types of structures in their own right, well-orderings. It makes no sense (apart from arbitrary stipulation) to call something an ordinal number except as part of a well-ordered system. The "absolute identity" of ordinals is as foreign to mathematics as is that of natural numbers or real numbers. But this principle goes further in two respects:

(i) it extends this attitude to sets generally, despite our tendency in ordinary discourse to speak of sets of objects as absolute (i.e. independent of their part in any richer mathematical structure). We can respect this tendency by distinguishing (as others have, cf., for example, Shapiro [1985]) between sets as members (or parts) of an iterative hierarchy, and classes as ordinary collections (if we cannot or will not avoid them entirely). This distinction will emerge naturally in a formal treatment.

(ii) The principle also rejects the notion of "the totality of all (possible) ordinals", in accordance with the ms approach to potential infinity. Thus, we could not respect the principle by taking ordinals as a proper class of (arbitrary) well-ordered urelements (as has been suggested, for example, by Menzel [1986]). Our only recourse is to take them as parts of structures, but what sort of structures should it be? Why not, one may ask, simply well-ordered systems, rather than models of set theory? Because the study of well-ordered systems is part of set theory, as key theorems (e.g. Hartogs')[8] teach us. (That is, the types of possible well-orderings depend on interactions among the set-theoretic axioms, e.g. Power Set and Replacement; they cannot be studied in full generality apart from such interactions.)

3. *Full classical power set operation*: There is no objection in principle to using second-order logic as a means of describing the structures.

Remark: This attitude is already implicit in our treatment of PA and RA. It is the classical operation of power sets that we are iterating when we come to set theory. The fact that there are basic ques-

---

[8] The (ZF) theorem of Hartogs states that for any set x, there is an ordinal not equinumerous with any subset of x. Replacement is used essentially in the proof.

tions concerning this operation that we do not know how to answer (and may never know how to answer) is not, for us, a reason to doubt that we have asked genuine questions.

4. *No proper classes*: There is to be no totality of "all possible structures", nor any union of the domains of all possible structures, etc.

Remark: As already indicated, this has to be squared with principle 3. A hint: note that in taking number theory to be about what would hold in any ω-sequence there might be, we are in no sense committed to "the totality of all possible ω-sequences". The latter is alien to the open-ended nature of mathematical construction, the recognition of which lies behind the rejection of proper classes.

5. *Extendability*: Every structure (of the first principle) has a proper extension, both in the sense of inclusion and in the sense that it, or some copy, occurs as a "member" of its proper extensions (i.e. in the domain of the relevant membership relation).

Remark: This is Zermelo's and Putnam's way of saying, in effect, that "the ordinals go on and on", and is closely related to the rejection of proper classes. Note, however, that it is stronger than that rejection: not only are there ordinals beyond those of any given structure, but there are structures of the relevant type as well. (This can be made to follow from principle 4, together with 2.) And, as we shall see, such extendability principles provide a natural way of generating many of the so-called "small large cardinals".

6. *Limited modality*: A primitive notion of logical possibility is employed, as in the ms treatment of PA and RA, in order to mark fundamental distinctions, principally that between quantifying over objects (Quine's '∃') and entertaining conceptual possibilities (Putnam's and our '◇∃', perhaps representing the "mathematical existence" of many "platonist" mathematicians). In the context of set theory, it is also employed to distinguish potential from actual infinity: inside a model, comprehension principles or other axioms guarantee infinite totalities ("actual infinities"); but possibilia are not recognized as objects, and we do not have any totality of "all possible structures", or ordinals, etc. Moreover, we do not quantify over possible worlds or intensions; we simply use modal operators.

Remarks: We are accustomed to giving set-theoretical semantics for modalities, and for a variety of logical purposes this is perfectly in order. But the msi of set theory, while aiming to respect such semantics as part of set theory, nevertheless, requires that its notion of logical possibility stand on its own. It functions as a primitive notion, and must not be thought of as requiring a set-theoretical semantics in order for it to be intelligible. Instead, of course, we may give modal axioms. (As discussed in Chapter 1, the natural choice of background logic would be S-5. To this, further modal-structural axioms must of course be added.)

It must be emphasized that the "counterfactuals" encountered in translations of ordinary mathematical sentences (e.g. of PA, RA, etc., and, as we shall see, of bounded set-theoretic sentences) involve a strict implication. In our formal semantics, we would not employ any relation of similarity among "worlds". And the "problem of contenability" afflicting ordinary and scientific counterfactuals (cf. Goodman [1955]) does not arise in these mathematical contexts. As the categoricity theorems demonstrate, all relevant conditions are explicitly given in the antecedents of the conditionals in question. Thus, there is no question as to what "further" background conditions need to be held fixed.

While we are on the topic of modality, let us consider the point raised early on concerning the alleged counter-intuitiveness of employing modality in connection with sets. The point was that modal-structural language in connection with sets seems strange since we conceive of sets as objects of a fixed, absolute kind. On this (standard platonist) view, sets do "have a nature" as objects, as it were (a thought that probably could be articulated only in a circular fashion, e.g. by saying that they are particular abstract objects which *contain* other objects, etc.). Thus, the objection runs, any attempt to modalize away our reference to sets automatically involves us in a distortion: sets cannot be just "points in a structure".

On the present approach, this appeal to "the nature of sets" is otiose. It may perhaps be traced to our conceiving of sets as determined by their members (extensionality): these—at the lowest level—are non-sets and presumably do "have a nature", or at least are taken as actual and are not to be modalized away as merely occupying positions in a structure. Perhaps we quite naturally transfer this attitude upward from elements to sets or collections.[9] But we can resist this temptation. Reflecting on ways of introducing set-language, we can view the very introduction of sets-as-objects-discourse as involving a reification of "results of selection activities,

---

[9] This was suggested to me by Howard Stein.

processes, or procedures".[10] Given objects of some type A and a selection procedure, p (or simply "selection" for short), we speak of a collection or a set as "the result of applying p to the A objects". And, correlatively, we introduce "membership" as holding between any A object selected by p and "the result". Modality can be understood as entering in two ways: first, we say: "given A objects and given selection procedure p (of some specified kind, say), *suppose there were* an object x correlated with p such that, for any A object, y, y belongs to x just in case p selects y. *Suppose*, further, that every selection p (of the specified kind) *gave rise* to such an object. Moreover, *suppose* that such objects *could* themselves serve as objects of further selections." (The emphases indicate the modality.) This already involves us in an iterative hierarchy, but it is restricted, implicitly, by the reference to specified kinds of selections. (A natural way of formalizing this would be through comprehension principles, the predicates themselves serving as a generalization of "selection procedure". This actually provides us with a way of justifying comprehension principles, as we shall see shortly.) The second use of modality is to free ourselves from the need to specify a selection procedure: we abstract from any restrictions on kinds of selection procedures and speak of all possible ways of selecting objects of the given type. When we combine this with the first stipulation, that "sets" may serve as "elements", we have, in effect, the operation of full power sets, the notion of possibly selecting (results of) all possible ways of selecting from among given objects. On this view, then, far from being foreign from, or antithetical to, our ordinary set-theoretic talk, modal idioms can be found, implicitly, at the very root of it. That sets are "permanent possibilities of selection" (as Putnam so aptly put it)[11] is more than just a picturesque phrase.

On this view, then, talk of the "existence of sets" is derivative. More fundamental is the idea of the possibility of selections, in terms of which existence of sets can be introduced. Let us, for the moment, not worry about attempting to recover a full system of set theory along these lines (something that would involve introducing axioms on "arbitrary selections"), and instead focus on comprehension principles, which could be used in more limited mathematical systems. Such comprehension principles can be arrived at as follows: let '$F(x)$'

[10] For an alternative treatment of set theory which also emphasizes selection activities, see Kitcher [1983].

[11] In his 'What is Mathematical Truth?' [1975], p. 71.

be a condition (possibly with parameters, and formulated in some specified language), and let 'S(p, x)' mean 'p selects x'. Then a typical set theoretic comprehension principle, of the form

$$\exists\alpha\forall x(x \in \alpha \equiv F(x)),$$

can be understood as introduced to *mean*

$$\Diamond\exists p\forall x(S(p, x) \equiv F(x)),$$

where p ranges over selection procedures (of some type that could be further specified, if desired).

Of course, 'F' in either scheme cannot be arbitrary in view of the Russell–Zermelo paradox. Moreover, this way of introducing comprehension principles applies most readily to predicative principles, in which the set quantifiers in the formula range over objects "already introduced", say at an earlier stage of a hierarchy. Whether such an approach can work for impredicative comprehension principles (such as that of second-order logic which we have employed throughout) is, I believe, an open question.

It should also be remarked that, if we wish to introduce talk of "possible results of selections" explicitly, we can by expanding the last formula to

$$\Diamond\exists p\forall x[(S(p, x) \equiv F(x)) \ \& \ \Diamond\exists\alpha(\alpha \text{ is the result of applying } p)].$$

Formally, it complicates things a bit to distinguish between possible selection procedures and their possible results. Philosophically, the distinction is important to this extent: we often deal with actual selection processes or procedures without having to speak of "results" in an abstract way appropriate for mathematics, i.e. as part of an iterative hierarchy. On the modal view, even in the case of actual selection processes, it is only necessary to entertain the *possibility* of such results. But then, when we come to entertain possible selections, the distinction between selections and results involves us in iterating $\Diamond$ s. It would be nice to avoid such complexities, by dropping the distinction. For mathematical purposes, it makes not the slightest difference whether we think of selecting results of selections or the selections themselves. Ordinary axiomatic set theory, of course, dispenses with the distinction, and its convenience is undeniable.

There are a number of immediate consequences of this way of looking at things, three of which are worth noting here. First, we obtain a natural justification of comprehension principles, provided we merely extend the notion of "selection procedure" to include the ordinary semantic relation of application of predicates (as suggested parenthetically a moment ago). For, in a comprehension principle, we are given a predicate, '$F(x)$'; and the principle can then be understood as merely asserting the possibility of using (applying) it; and such use counts as a witness to the '$\Diamond \exists p$' in '$\Diamond \exists p \forall x[S(p, x) \equiv F(x)]$'. In other words, we understand the S relation in such a way as to guarantee each such instance of comprehension simply on the basis of a Tarski biconditional:

$$\forall x(\text{'F' applies to } x \equiv F(x)),$$

which is itself readily justified on the basis of linguistic practice. (We needn't insist that it be "analytic".)

A second consequence is that reference to the null set is made to disappear entirely. For

$$\exists \alpha \forall x(x \notin \alpha),$$

guaranteed by taking 'F' as, say, '$x \neq x$', comes to nothing more than

$$\Diamond \exists p \forall x(\sim S(p, x)),$$

i.e. there could be a procedure which selects nothing, something we can be quite confident in by simply adopting a posture of utter lassitude (preferring this perhaps to "honest toil"). (To obtain uniqueness of the null set, further assumptions would be needed, such as an axiom guaranteeing that every selection procedure corresponds to a unique possible selection result.)[12]

---

[12] One way of formulating such an axiom is to use two relation terms, $S(p, x)$ and $\in(x, y)$ (which we rewrite as '$x \in y$', as usual), and modal quantifiers. Then, to express the possibility of "sets" ("results of selections") uniquely correlated with selections, we could write, as a "fundamental correlation axiom",

$$\Box \forall p \Diamond \exists! \alpha_p \Box \forall x(S(p, x) \equiv x \in \alpha_p), \tag{FCA}$$

in which $\exists!$ abbreviates a uniqueness clause in the usual way, i.e. FCA fully written out becomes,

$$\Box \forall p \Diamond \exists \alpha_p \Box \forall x[(S(p, x) \equiv x \in \alpha_p) \ \& \ \Box \forall \beta \{\Box \forall y(S(p, y) \equiv y \in \beta) \supset \alpha_p = \beta\}].$$

(Reasoning with modal quantifiers is exactly analogous to ordinary quantifier reasoning, as codified, e.g. in Cocchiarella [1966]. We would prefer to call Cocchiarella's system a "logic of actual and modal quantification".)

An immediate consequence is extensionality, in the form,

$$\Box \forall p \Box \forall q[\Box \forall z(S(p, z) \equiv S(q, z)) \supset \alpha_p = \alpha_q],$$

whence,

$$\Box \forall y((y \in \alpha_p \equiv y \in \alpha_q) \supset \alpha_p = \alpha_q),$$

which expresses extensionality for sets (i.e. the selection results). Further axioms for sets can be obtained from suitable axioms governing S. From the standpoint to be

The last consequence we shall mention is that we readily account for "exceptional claims" frequently made concerning the possibility and necessity of "sets", such claims as,

> If it is possible that sets exist, then they actually exist,

and,

> If sets actually exist, then they necessarily exist,

claims which are exceptional in that they involve a collapse of all modal distinctions for "objects of this sort" (whereas one may still uphold those distinctions "in the usual sort of case"). However, if '∃' for sets is understood as introduced via '◇∃' for selections, these statements are immediate consequences of the S-5 modal axioms (assuming "x actually exists" is understood simply as, say, '$\exists y(y = x)$'). If we write '$S'(p)$' to abbreviate the statement that "p selects something or is a selection of nothing" (i.e. p enters into the S relation), then the first claim can be understood as taking on the form,

$$\Diamond \Diamond \exists p(S'(p)) \supset \Diamond \exists p(S'(p)),$$

and the second as,

$$\Diamond \exists p(S'(p)) \supset \Box \Diamond \exists p(S'(p)),$$

which are instances of the characteristic S-4 and S-5 modal axioms, respectively. (This sort of result could even be taken as a kind of "confirmation" of our choice of S-5 as the background modal logic. Cf. Chapter 1, n. 8.)

Given this modal perspective, one could, at this point, attempt to arrive at a full set of axioms for set theory by trying to formalize further these ideas concerning "possibilities of selecting". This could proceed in various directions. By suitably restricting the notion of "selection procedure", one could build various modal systems of (more or less) constructive set theory. But, from our non-constructive point of view, we would be trying to axiomatize a notion of "arbitrary selection", and, very likely we would find ourselves

developed below, this procedure for arriving at set theory encounters an awkwardness in connection with the use of unrestricted modal quantification over selections, since an intended model of such axioms would presumably be inextendable (i.e. containing "all possible selections or sets"). Thus, a careful axiomatization along these lines should seek to restrict the FCA in some natural way. One advantage of our direct, structural approach, to be carried out below, is that we can express extendability principles explicitly and can avoid commitment to inextendable models.

mimicking standard axiomatic set theories (substituting $\lozenge \exists$ for $\exists$, of course, and referring to "selection of something" or "result of selection of something", in place of "non-empty set"). (Such axioms could be added to the fundamental correlation axiom of n. 12, or to something along those lines.) Instead, there is a more direct approach. For, as we shall see, once we admit the use of modality and the framework of second-order logic, we can carry out a modal-structural interpretation by appealing to already well-worked-out axiom systems, in much the manner of the msi of arithmetic and analysis. And we may justify our choices of axiom systems by appealing to their utility in actual mathematical practice and to the fact that they can be used to characterize the types of structures of actual mathematical interest. Indeed, further questions of justification—especially of modal existence postulates—will remain, but these are best taken up case by case as they arise.

These, then, are our guiding principles. Let us now see how they can be given a more precise formal expression and how they may be put to use.

§ 2. The Relevant Structures

It is taken for granted, at the outset, that the ZF axioms are "motivated" in the sense that they attempt to spell out—if only partially—a coherent conception of "iterative hierarchy". This conception (as emphasized by Gödel, Tait, and others)[13] is based on the joining together of two ideas, the idea of "forming all possible collections" from a given collection of objects—no matter what they may be—, and the idea of iterating this "operation" into the transfinite. It is striking that one may develop axioms designed to express these ideas in a variety of ways that initially appear to be quite different but which in fact lead to the same result. Thus, in addition to the familiar ZF axioms, which can be obtained from assumptions on "stages" (as in Boolos [1971], cf. Shoenfield [1977]), there is also the system developed by Scott based on cumulative levels (see Scott [1974]), which turns out to be equivalent to ZF. (The striking result here was that well-foundedness followed from what appear to be "more basic" axioms on levels.) These results strongly suggest that we somehow

---

[13] See Gödel [1947], pp. 476–7, whose attitude toward strong axioms of infinity is supported here; also, see Tait [forthcoming].

make use of one or another of these axiom systems in attempting to characterize "structures of the appropriate type" that can serve as "the subject-matter" of set theory on the iterative conception.[14]

To see how this may be done, recall (from Chapter 1) the ms treatment of number systems, e.g. the natural numbers. Following Dedekind, we used his second-order axioms (of Peano Arithmetic) to define "ω-sequence":

$$(X, f) \text{ is an } \omega\text{-sequence iff } (\wedge \text{PA}^2)^X \binom{s}{f}, \tag{2.1}$$

where the right side abbreviates the result of writing out the PA^2 axioms with all quantifiers relativized to the second-order variable X, replacing each occurrence of the successor symbol 's' with the two-place relation variable 'f'. This led directly to a ms interpretation of arithmetic in which ordinary arithmetic sentences were represented as saying what would hold in any ω-sequence. We then appealed to the categoricity of the PA^2 axioms in order to establish certain desirable properties of this representation.

Something analogous can be done for set theory. If we employ second-order logic, we can replace the infinitely many first-order Replacement axioms with a single second-order Replacement Axiom (as in Zermelo, [1930]), obtaining a finite list, ZF^2. We may then explicitly define "natural model (of set theory)" along the lines of (2.1):

$$(X, f) \text{ is a natural model iff } (\wedge \text{ZF}^2)^X \binom{\in}{f}, \tag{2.2}$$

in which, on the right, the two-place relation variable, 'f', replaces '$\in$' throughout. We may then attempt to interpret set theory as the study of structures of this type.

What justifies the terminology, "natural model"? The answer to this question is central to the interpretation and will, at the same

---

[14] There need be no claim that this conception is the only coherent one meriting the name "set theory". For example, mathematicians may be interested in investigating non-well-founded structures, or structures with ultimate classes, and so on. It is an advantage of the structuralist approach that it can, in principle, accommodate all of these theories. No contradiction arises precisely because the axioms of these systems are not viewed as categorical assertions, true or false by themselves, but rather as (perhaps only partially) characterizing a possible type of structures that one wishes to investigate. It is only the possibility of the structures in question that need be categorically asserted.

time, explain why the choice of ZF^2 is mathematically motivated and not simply a matter of technical convenience (i.e. to obtain a finite list so that a definition like (2.2) can be presented).

As already indicated, the key technical results proved by Zermelo [1930] showed that full models of the ZF^2 axioms are essentially determined (i.e. up to isomorphism) by two parameters: the "width" (cardinality) of the urelement basis and the ordinal height. If urelements (besides the null element) are not allowed (i.e. extensionality is unrestricted, as in the usual presentations of the axioms), it follows from Zermelo's theorems that, for any two full[15] models, M_1 and M_2, either M_1 is isomorphic to an end extension of M_2 or M_2 is isomorphic to an end-extension of M_1 (where "M is an end extension of N" means that (i) the domain of N is included in the domain of M and $\in^N$ is the restriction to N of $\in^M$ (i.e. M is an extension of N), and (ii) for any x in N, if $y \in^M x$, then $y \in^N x$). (We may summarize this result by saying that ZF^2 is *quasicategorical*.) It follows, moreover, that every such full model is isomorphic to one of the form $(V_\kappa, V_{\kappa+1}, \in| V_{\kappa+1})$, where κ is a strongly inaccessible cardinal. (Here V_κ is defined in set theory as the κ'th cumulative level (also denoted R_κ), i.e.

$$R_\alpha = \bigcup_{\beta < \alpha} \mathscr{P}(R_\beta),$$

where $\mathscr{P}$ is the power set operation; the second item of the model, $V_{\kappa+1}$, gives the range of the second-order quantifiers occurring in the ZF^2 language. To simplify notation, we shall sometimes denote such models merely by 'V_κ'.)

It is precisely models of this form that are called "natural models" in set theory. If one speaks in the usual way of a real world of sets, the membership relation of these models is just the "real membership relation", and the models V_κ are thought of as simply initial segments of the real universe of sets (i.e. pure sets, beginning with just the null set). Hence the term "natural". Note that, unlike the universe, the domain of any natural model is a set, and any such model itself can be treated as a set, i.e. as in the range of the first-order quantifiers of set theory. Thus, as standardly presented, Zermelo's theorems are

---

[15] A model is called *full* provided (i) its power set operation is maximal at every stage (i.e. any proper extension must contain more ranks); and (ii) the ranges of the second-order n-ary relation quantifiers ($n = 1, 2, \ldots$) are full, i.e. each such range is the set of all subsets of $|M|^n$, where $|M|$ is the domain of the model M.

theorems of set theory, and in fact of the first-order ZF axioms.[16] In effect, second-order logic is mentioned—when we speak of models of the ZF^2 axioms—but it need not be used.

It should be stressed, however, that this role of second-order Replacement is critical. It is a striking fact that natural models of the ZF^1 axioms—i.e. of the form $(V_\alpha, \in|V_\alpha)$—need *not* have inaccessible height. (The height can be a singular cardinal, as can be proved by a Löwenheim–Skolem type construction.)[17] This is but another symptom of the weaker expressive power of first-order axioms as compared with second order, analogous to the better-known fact that models of the first-order axioms, including the Regularity Axiom, need not even be well-founded (i.e. they may contain infinite descending $\in$-chains). (In the second-order case, Replacement succeeds in ruling out such non-standard models.)

Now the crucial point to observe, in the interests of a structuralist interpretation, is that the quasi-categoricity of ZF^2 can be suitably stated and proved entirely within the pure second-order logical framework, much as the full categoricity of PA^2 and that of RA^2 were recoverable from within that framework. The statement takes the following form:

Quasicategoricity of ZF^2:

$$\forall X \forall f \forall Y \forall g \left[(\wedge ZF^2)^X \binom{\in}{f} \& (\wedge ZF^2)^Y \binom{\in}{g} \supset. \right.$$

either $\exists \phi (\phi$ maps X 1–1 on to a substructure $(Y', g| Y')$ of (Y, g) such that

(i) $g(\phi(x), \phi(y))$ for any x, y such that $X(x) \& X(y) \& f(x, y)$, and

(ii) (Y, g) is an end extension of $(Y', g| Y'))$;

or $\exists \phi (\phi$ maps Y 1–1 on to a substructure $(X', f| X')$ of (X, f) such that etc. (i.e. clauses just like (i) and (ii) but with the roles of X, f and Y, g reversed))].

It is an exercise to write this out in detail as a second-order

---

[16] See e.g. Lévy [1960], who formulates similar theorems in terms of a ZF^1-definable notion of "standard complete model". Cf. Drake [1974], pp. 122–3. A number of results of this type were obtained, apparently independently of Zermelo, by Sheperdson [1951], cited by Lévy.

[17] See Drake [1974], p. 111, Ex. 1.6(4).

statement. The proof also can be given in axiomatic second-order logic.[18]

Several remarks are in order concerning the significance of quasi-categoricity, especially in this second-order form.

First, it should be noted that the talk of "structures" in the theorem and its proof is abbreviatory of direct second-order statements (on the plan of '$(\wedge ZF^2)^X\binom{\in}{f}$' as short for '$(X,f)$ satisfies ZF^2'); a set-theoretic definition of satisfaction is not involved. The reasoning takes place entirely within second-order logic, just as in the recovery of the categoricity of PA^2 and RA^2. Thus, we do not *have* to think of these theorems as taking place either within a larger model of set theory or within a fixed universe of "all sets". This is significant, in that the first view encourages a kind of relativism which may be unwarranted—i.e. the "background model" might be non-standard—and the second view encourages acceptance of proper classes. The second-order formulations attempt to get by without either.

In this connection, note that, since we do not need to speak literally of model-theoretic structures, we do not need to invoke explicitly the distinction between full and non-full structures (i.e. between structures in which second-order (k-ary) quantifiers range over all subsets of (the k-fold Cartesian product of) the domain, and those with smaller ranges for the second-order quantifiers). Instead, we simply reason directly with universal second-order statements, e.g. Replacement, and this suffices. (The slogan would be, "'All' means *ALL*.")

Secondly, provided we accept the second-order logical axioms,

---

[18] One way to proceed is to prove first that the ordinals of X (say, the von Neumann ordinals, in the sense of f) are order-isomorphic (under a map ψ) to an initial segment of the ordinals of Y (in the sense of g), or vice versa. This uses just second-order comprehension and transfinite induction inside X and Y, given by the hypothesis of the theorem. Next one extends one of the maps (ψ or ψ^{-1}) to a map ϕ on the Rα of the structures, ensuring that "membership is preserved", i.e. if, say, ϕ is from X into Y and the rank of x is α, $\phi(x)$ will be that object y of Y of rank $\psi(\alpha)$ such that any y' of Y bears g to y just in case it is $\phi(x')$ for some x' bearing f to x. The properties of the theorem then follow. (In proving that ϕ (say it is from X into Y) is onto the Rα (in the sense of g) of a segment of Y, second-order Replacement is used. If ϕ were not onto, there would be a descending g-chain of objects, each outside the range of ϕ; by construction of ϕ, for each non-empty (in the sense of g) item of the chain there must be a further item of lower rank. By well-foundedness of the model (guaranteed by Replacement^2), the chain must be finite; but then, it must terminate in the null set of Y, $\varnothing^Y = \phi(\varnothing^X)$, which is a contradiction.)

quasicategoricity supports the conclusion that the ZF^2 axioms do succeed in determining a definite type of structures, given by the definition (2.2) ("natural models"). Once the urelement basis is fixed, only the height is left open (i.e. up to isomorphism). As an immediate corollary, it follows that all sentences of ZF with bounded quantifiers (i.e. restricted to sets below a given rank) are determined, i.e. their truth value is fixed identically in all natural models of sufficient height. Such sentences may be undecidable from the first-order axioms, but, if one believes the quasicategoricity theorem, they are determinate none the less. (Thus, e.g., the Continuum Hypothesis is determined already at $V_{\omega+2}$, if a flat pairing function is used. This isn't yet a natural model of ZF^2 of course; but it follows that all natural models of ZF^2 (which includes the Axiom of Infinity) agree on CH. Similarly, with Souslin's Hypothesis and all undecidables pertaining to accessible levels.)

Now it may be (and has been) argued that quasicategoricity and the use of second-order logic are really irrelevant to this determinateness (cf. Weston, [1976]). Determinateness, if guaranteed by anything, is guaranteed by our ordinary understanding of the set-theoretic sentences themselves (in the interesting cases of CH, Souslin, etc., first-order sentences), i.e. our understanding of sets and membership. Moreover, such understanding is involved in the quasi-categoricity proof itself (as ordinarily presented, that is, in our reasoning concerning full models of ZF^2), and hence cannot be relied upon even to bolster our faith.

However, on the structuralist view, the situation is different. The ordinary (bounded) set-theoretic sentences need not be interpreted over a unique cumulative hierarchy. Rather they may be understood elliptically as saying what would be the case in any sufficiently high natural model, on exactly the plan of the modal-structural interpretation of number theory and real analysis, i.e. each bounded sentence S, of rank ρ of $\mathcal{L}(ZF)$ (either ZF^1 or ZF^2) could be translated as

$$\Box \forall X \forall f \left[(\wedge ZF^2)^X \binom{\in}{f} \,\&\, X(\rho) \supset S^X \binom{\in}{f} \right].^{19} \tag{2.3}$$

---

[19] If ρ can be proved to exist in ZF (as it typically can in cases of normal mathematical interest), the clause '$X(\rho)$' is redundant. Otherwise, if ρ is provably unique in fulfilling a condition $C(x)$, '$X(\rho)$' becomes '$\exists x(X(x) \,\&\, C(x))$'. Actually, it suffices to require $\forall \beta < \rho\, X(\beta)$, but because we quantify only over ordinals in models (cf. our informal requirement 2 of §1), this becomes the more complicated statement that X contains (up to isomorphism) every β of any ZF^2 model, (N, g), such that N contains an ordinal ρ such that $C(\rho)$ and $\beta < \rho$ ($<$ in the sense induced by g).

Now quasicategoricity takes on a new importance, for it would be used to argue that the translation pattern (2.3) is fully bivalent (in direct analogy with the msi of PA and RA, as in Chapter 1). Moreover, as just indicated, the argument does not rely on embedding the structures in a further model of set theory, but, rather, on the far more modest second-order logical axioms. And, of course, no assumption of a unique cumulative hierarchy is involved.[20]

Thirdly, the present formulation of quasicategoricity and the associated translation scheme (2.3) for bounded sentences may be regarded as formalizing the distinction between "logical class" and "iterative set" (cf. Shapiro [1985]): logical classes and relations are (an interpretation of) what the second-order logical quantifiers range over, whereas iterative sets are construed structurally, along with ordinals, as arbitrary items interrelated in the right ways.

However, "logical classes" may not be the only way of reading the second-order notation. In addition to Fregean concepts and various intensional interpretation (cf. the work of Cocchiarella, in particular [1986]), there is even the possibility of "nominalistic" interpretations (as suggested in Putnam [1967]). The natural models could be thought of as "concrete" (though probably not as concrete (an overly concrete special case?)), i.e. characterized by means of the calculus of individuals together with "nominalistically acceptable" predicates. We shall not pursue such interpretations here, but will touch on the question again in Chapter 3.

Having seen how to talk about natural models, we are now in a position to write out more precisely the modal-existence postulates embodied in the informal principles of § 1, and to specify more precisely how modality enters in resolving the tension between the use of second-order logic (principle 3) and the rejection of proper classes (principle 4). Concerning modal existence, there is, first of all, the assumption that natural models are possible, which can be written simply as

$$\Diamond \exists X \exists f (\wedge ZF^2)^X \binom{\in}{f}. \tag{2.4}$$

This functions as a working hypothesis of realist set-theoretic practice, just as the possibility of ω-sequences or of separable ordered continua function, respectively, in the practice of number theory and

---

[20] Weston [1976] attributes such an assumption to Kreisel [1967], whose invocation of quasicategoricity in support of determinateness he criticizes.

real analysis. However, (2.4) is much less directly tied to experience than these latter modal-existence assumptions, and is, in this respect, far more speculative. How best to describe and assess our "evidence" for such a hypothesis remains one of the most difficult challenges confronting mathematical epistemology.[21] Here we are concentrating, however, on the prior task of articulating basic assumptions and exploring their implications.

The second modal-existence assumption is the *Extendability Principle* (EP). Informally stated, "Every natural model has a proper extension." This can be written in modal second-order logic as

$$\Box \forall X \forall f \left[(\wedge \mathrm{ZF}^2)^X \binom{\in}{f} \supset \Diamond \exists Y \exists g \left\{ (\wedge \mathrm{ZF}^2)^Y \binom{\in}{g} \right.\right.$$
$$\left.\left. \& (X,f) < (Y,g) \right\} \right] \quad (2.5)$$

in which the last conjunct abbreviates the (second-order) statement that X is a proper subclass of Y and f is the restriction of g to X.[22]

As in the treatment of PA and RA, our formal apparatus is second-order S-5. However, it is essential that the comprehension scheme of second-order logic be restricted to *actualist instances*, i.e. those in which the initial universal quantifier is simply $\forall$, not $\Box\forall$. We do *not* allow, for example,

$$\exists Y \Box \forall y \left[Y(y) \equiv \Diamond \exists X \exists f \left((\wedge \mathrm{ZF}^2)^X \binom{\in}{f} \& X(y) \right) \right],$$

that there exists a class of all elements of any possible model.[23] Such a beast would be the modal-structuralist counterpart of a proper class and would clearly violate the spirit (and, given other assumptions, even the letter) of the EP. Such conflicts can be avoided by restricting comprehension in the following ways: (i) only actualist

---

[21] For a recent study of modal interpretations of set theory which supports the conclusion that the mathematical modality of a wide class of such interpretations is not reducible to the notions of "naturalistic semantics", see McCarthy [1986]. This accords with our own view throughout, that the characteristic modal existence postulates of the msi cannot be understood as "analytic" in any useful sense.

[22] Note that we cannot express directly in second-order notation that X or f occur as "elements" of Y (i.e. we cannot write $Y(X)$ or $Y(f)$). However, we can state (and prove) that X and f are equivalent to elements of Y, e.g. that $\exists y(Y(y) \& \forall z(X(z) \equiv g(z, y))$, etc. For mathematical purposes, this seems to suffice.

[23] Such stronger forms of "possibilist comprehension" (for modal as well as non-modal formulas) are characteristic of some of the higher-order modal logics investigated by Gallen [1975] and Cocchiarella [forthcoming].

initial '∀' is permitted, as just explained; and (ii) only instances involving non-modal formulas are taken as axioms. (The above example is thus ruled out on two counts.) The first restriction is the key to avoiding commitment to proper classes while still employing second-order logic. The second enables us to keep quantification into modal contexts to a minimum. (For our purposes, there appears to be no need to invoke comprehension with modal formulas.)

We do, of course, have the straightforward necessitation of (actualist, non-modal) instances of comprehension. Thus, we may continue to speak of proper classes relative to a model. For example, in the context of a given model, its domain or the class of its ordinals, etc., are recognized, and they behave like proper classes, relative to the model. But, by the EP, they are represented as elements of a further model, i.e. they can be treated "as sets". But we never have an object corresponding to the "totality of all possible models", or "the totality of all possible ordinals", etc., much as, in the case of arithmetic, we never countenance "the totality of all possible ω-sequences". Such totalities are regarded as incoherent, since any totality—treated as an object—can be extended.

§3. Unbounded Sentences: Putnam Semantics

We have seen that bounded sentences (of $\mathscr{L}(ZF^2)$) can be interpreted along modal-structuralist lines in essentially the manner of the ms treatment of arithmetic and analysis. Now it should be clear that, once we recognize the possibility of essentially different models—as we do once we have adopted the EP—such a simple translation pattern will not work for unbounded sentences. For example, let $P(x)$ represent a large cardinal property (e.g. "x is strongly inaccessible"), and consider the sentence, $\exists\kappa(P(\kappa))$. Obviously, we cannot treat this as saying that in any possible (full, well-founded) ZF^2 model, there would be an object κ such that $P(\kappa)$, for, if κ' is the least κ such that $P(\kappa)$, then (if P implies strong inaccessibility) the class of sets of rank κ' (speaking set-theoretically) would constitute a model lacking any P. Thus, different models give different answers and the simple translation pattern in question would treat both $\exists\kappa(P(\kappa))$ and its negation as false.

It should be pointed out that, of course, there are ways of restricting the heights of models uniquely so that the simple translation pattern would work. One could insist, for example, that there be no inaccessibles (i.e. one would add this statement as an axiom to ZF^2); the resulting theory can be proved categorical, so that the simple translation scheme would be fully bivalent.[24] However, there are (at least) two substantial objections to this course. The first would be the *arbitrariness* of the Axiom of Restriction. If the aim is to produce a categorical theory, one can achieve this in infinitely many different ways (e.g. add to ZF^2 the axiom that there are exactly 17 inaccessibles, etc.), and no evident reason to single out one as optimal. But second, and more fundamentally, such axioms conflict with the EP which expresses something too deeply rooted in our use of set-like operations to renounce, the possibility of "going beyond" any definite totality.[25] We must try to live with Extendability.

Having made this decision, we may attempt to interpret unbounded set-theoretic sentences as saying what would hold in relevant models and their extensions. Consider, for example, a simple AE sentence, say, $\forall\alpha\exists\beta(\beta > \alpha \,\&\, \text{Inac}(\beta))$. We may construe this as saying that, for any natural model (X, f) and any ordinal α in X, it is possible that there is a (not necessarily proper) extension (Y, g) of (X, f) with a β in Y such that $\beta > \alpha \,\&\, \text{Inac}(\beta)$. (Here, α, β range over ordinals. 'α in X' would of course be written $X(\alpha)$. Second-order variables can be treated by writing out $Z \subseteq X$, etc. We may employ "in X" ambiguously to cover both cases.) One may carry out this plan for arbitrary, unbounded sentences, first putting the sentence into prenex form and then iterating the appropriate modal quantification over extensions, for each quantifier of the prefix. If we use M, N, etc. as special variables ranging over natural models (i.e. pairs (X, f), etc., treated according to the definition of 'natural model' above), then a sentence of the form, say,

$$\forall x\exists y\forall z A(x, y, z),$$

would be translated as

$$\Box\forall M\forall x[x \text{ in } M \supset \Diamond\exists M'\exists y\{y \text{ in } M' \,\&\, M \leq M'$$

$$\&\, \Box\forall M''\forall z(z \text{ in } M'' \,\&\, \ M' \leq M'' \supset A(x, y, z))\}].$$

(Here the variables x, y, z may be first or second order, and '$\leq$'

[24] Cf. Fraenkel's equivalent Axiom of Restriction, in Fraenkel, Bar-Hillel, and Lévy [1973], pp. 114 ff.
[25] See Fraenkel, Bar-Hillel, and Lévy [1973], p. 118, for an expression of essentially this point, despite the authors' willingness to speak of "the universe".

abbreviates the second-order conditions for "is a submodel of".) It should be clear from this example how to generate the Putnam modal translate of an arbitrary sentence S of $\mathscr{L}(ZF^2)$. We will refer to it as S_{msi} ("S on the modal-structural interpretation").[26]

What are we to make of this translation scheme? Surely, by itself, it provides us with no independent handle on unbounded set-theoretic truth, for the translates contain unbounded modal quantifiers "over arbitrary extension of models". Without further assumptions—beyond the ZF axioms—on what sorts of models are possible, there are many questions that we cannot expect to answer. However, it should be stressed that even so elementary an assumption as the EP can have important consequences. As a cardinal example, consider the Axiom of Inaccessibles,

$$\forall\alpha\exists\beta(\beta > \alpha \,\&\, \text{Inac}(\beta)). \tag{AI}$$

Evidently, its modal-structural translate, $(AI)_{msi'}$ is implied by our assumptions, the EP plus second-order logic. (Any natural model, M, is of the form V_κ, κ strongly inaccessible; this is proved in essentially the manner of Zermelo [1930]; cf. also Drake [1974], p. 112. The proof can be adapted to the present setting by carrying it out relative to any proper extension of M (which must contain κ).) Thus, this large cardinal axiom appears, not as an arbitrary extra assumption, nor by invoking dubious totalities (e.g. by reflection on "the universe" (cf. below, § 5)), but as a direct consequence of a structural principle motivated by our understanding of set-like operations. How such methods may be further extended will be taken up in § 4.

Although our access to the modal translates depends on strengthenings of ZF, still we may ask for comparisons between the msi and the usual fixed universe picture. It is not difficult to show that, *from within the latter point of view*, there is complete agreement between the two with respect to all first-order questions of ZF, decidable or not.

Working within set theory (eventually, it will have to be at least as strong as NGB in order to speak explicitly of "the universe"), it is possible to generate the Putnam translation scheme from a simple set of semantical rules for evaluating sentences of the (non-modal) set-theoretic language ($\mathscr{L}(ZF^2)$) pointwise in the intended

---

[26] Essentially this translation pattern was originally proposed by Putnam [1967]. We have cast his informal exposition in second-order logical terms and have shifted from his "standard concrete models of Z" to "natural models of ZF".

structures, i.e. in full, well-founded models of ZF^2, isomorphic to the natural models (in the usual terminology), V_κ, with κ inaccessible. In presenting these rules, we make the simplifying assumption that the modal operators are absorbed into the quantifiers with which they are uniquely associated in the translates (viz. '$\Diamond\exists$' becomes '$\exists$' and '$\Box\forall$' becomes '$\forall$'). Letting M, M', etc. range over full ZF^2 models, as above, and letting E, E', etc. range over evaluations of variables, we can introduce a new satisfaction relation, $M \vDash_P A[E]$, "M Putnam-satisfies A at E", for A a formula of $\mathscr{L}(ZF^2)$, inductively, as follows:

1. A atomic: $M \vDash_P A(x, y)[E]$ iff $M \vDash A(x, y)[E]$, where $\vDash$ is ordinary satisfaction;
2. $M \vDash_P \sim A[E]$ iff $M \nvDash_P A[E]$;
3. $M \vDash_P A \mathbin{\&} B[E]$ iff $M \vDash_P A[E] \mathbin{\&} M \vDash_P B[E]$;
4. $M \vDash_P \exists u A[E]$ iff $\exists M' \geq M \exists E'$ agreeing with E except possibly at u such that $M' \vDash_P A[E']$.

'$\forall$' is introduced classically in the usual way as $\sim\exists\sim$. Using this definition, if one simply reads off the relevant quantifier clauses and applies disquotation (substituting quantification over items in the structures for quantification over evaluations of variables), one recovers the Putnam translate of A (modulo the absorption of modal operators into quantifiers). Note that Putnam semantics is just like ordinary semantics except for the quantifier clause in which '$\exists$' is understood as $\exists$ with respect to some (possible) extension of the structure in question.

How do different intended models compare with respect to Putnam semantics? One might imagine that, as one proceeds outward along a path of such models, the truth value of a ZF sentence could shift, as it can in the case of ordinary semantics. However, it is easy to show that this cannot happen. Using the Zermelo quasicategoricity theorems, there is (up to isomorphism) only a single path of relevant models (they can be identified with the natural models V_κ, κ inaccessible), and one has a

Stability Theorem: Let A be a sentence of $\mathscr{L}(ZF^2)$ and M a full ZF^2 model;

then $M \vDash_P A$ iff $\forall M' \geq M M' \vDash_P A$.

The proof is by induction on the number of quantifiers in A. (See Appendix.) This immediately yields as a

Corollary: All full ZF^2 models are elementarily equivalent with respect to Putnam semantics, i.e. for any sentence A of $\mathscr{L}(ZF^2)$ and for any such models M, N,

$$M \vDash_P A \text{ iff } N \vDash_P A.$$

(Proof: Let $M \vDash_P A$. By the Zermelo quasicategoricity theorems, we may without loss assume that either $M \leq N$ or $N \leq M$. In the first case, $N \vDash_P A$ by the stability theorem; in the second, if $N \nvDash_P A$, $N \vDash_P \sim A$ and, again by the theorem, $M \vDash_P \sim A$, contradiction.)

Thus, Putnam semantics gives unique answers to all (ZF) set-theoretic questions, regardless of the point of evaluation. But does it give good answers? Yes, in this sense: it gives exactly the answers to all first-order questions that the fixed set-theoretic universe gives, assuming the Axiom of Inaccessibles. That is, we have a

Correctness Theorem: Let $A(\mathbf{a})$ be a sentence of $\mathscr{L}(ZF^1)$ with parameters $\mathbf{a}$ ($= a_1 \ldots a_n$) and suppose $V \vDash A(\mathbf{a})$; then $\exists \kappa$ such that V_κ is a full ZF^2 model, the a_i are in its domain, and $V_\kappa \vDash_P A(\mathbf{a})$.

The proof again is by induction on the number of quantifiers in A (see Appendix). Note that, by stability, this is equivalent to proving, from the same hypotheses, that $\forall \kappa$ such that V_κ is a ZF^2 model containing the a_i, $V_\kappa \vDash_P A(\mathbf{a})$. The converses then follow by the clauses for negation.

Note that the restriction here to first-order sentences is essential. For example, V satisfies the second-order sentence, $\exists X \forall \alpha (\alpha \in X)$, but no V_κ can Putnam-satisfy this, in light of *Extendability* (equivalently, the Axiom of Inaccessibles). Although $\vDash_P$ is defined for second-order as well as first-order sentences, it "agrees with the universe" in general only on the latter. As the example just given illustrates, a rejection of proper classes is built in (as it should be, given the viewpoint for which it is designed). But, in light of *Stability*, which applies to both first- and second-order sentences, the rejection is "once and for all".

It must be emphasized that the "correctness theorem" depends critically on the Axiom of Inaccessibles as this is understood on the fixed universe view: the inaccessibles, hence natural models V_κ, must be cofinal in the ordinals. If this assumption is relaxed—if, for example, we step outside the fixed universe view and do not even allow talk of a structure with "all ordinals"—then the argument breaks down. And, in that case, further assumptions governing the possibilities of full models will be needed even in order to guarantee that all instances of the ZF^1 axioms hold on $\vDash_P$ (at some or any full model). We have the following curious situation: although the simple EP suffices to guarantee that

$$\vDash_P \forall\alpha\exists\beta(\beta > \alpha \ \& \ \text{Inac}(\beta)),$$

i.e. the Axiom of Inaccessibles holds on $\vDash_P$, and although

$$\vDash_P \text{Replacement}^2,$$

i.e. the second-order Replacement axiom holds on $\vDash_P$, stronger extendability principles than the simple EP are needed to guarantee that

$$\vDash_P \text{Replacement}^1,$$

for all instances of the first-order scheme. To see this, first observe that the second-order axiom does hold on $\vDash_P$. This axiom is

$$\forall F \forall a [\forall x \forall y \forall z (F(x, y) \ \& \ F(x, z) \supset y = z) \supset$$
$$\exists b \forall v (v \in b \equiv \exists u (u \in a \ \& \ F(u, v)))].$$

In evaluating it on $\vDash_P$, at any given M, we look at arbitrary extensions N of M and any $F \subseteq$ *the domain of* N fulfilling the antecedent. Since we have this restriction on F, further extensions of N brought in by the remaining quantifiers play no role. The consequent is thus guaranteed by the assumption that N is a model of ZF^2 (i.e. if $F(u, v)$, u and v must both already be in N).

However, suppose $\forall F$ is dropped and 'F' is replaced by a formula, $\phi(x, y)$, with two free variables, i.e. we consider an arbitrary first-order Replacement axiom. Then there is no built-in restriction of the function defined by the formula ϕ to the domain of any model, and further extensions brought in by the quantifiers may be relevant. Consider, for example, the formula $\psi(i, \kappa)$ which says that κ is the ith inaccessible, for $i \in \omega$. If the possible full models are assumed to form merely an ω-sequence rather than a transfinite one, i.e. all possibilities are represented by

$$V_{\kappa_1} < V_{\kappa_2} \ldots < V_{\kappa_j} < \ldots \qquad (j < \omega),$$

then there is no range of ψ on ω. But this is compatible with the simple EP. In order to rectify the situation, it is necessary to assume some stronger extendability principle.[27]

One such principle which suffices is as follows:

> Let $\phi(x, y)$ be a formula "defining a function", where this is spelled out by writing out the Putnam translate of the usual condition; further let a be any set in any full model such that, for any x in a, M_β is the least full model containing the unique y such that $\phi(x, y)$. Then it is possible that there exists a common proper extension, M, of all such M_β. (Strong EP)

Now it can be shown that all instances of Replacement[1] hold on $\vDash_P$. (I see no obstacle to formalizing the proof in axiomatic second-order S-5.) And it may be expected that still stronger extendability principles will be needed to verify stronger conditions on height.

§4. Axioms of Infinity: Looking back

We have already seen how (the possibility of) endlessly many inaccessibles can be generated from basic assumptions on natural models, framed within modal second-order logic. Before considering further ascent, we should look back down and ask about the very first axiom of infinity, the one that merely asserts the existence of some infinite set. Recall that this is independent of the other axioms of ZF. ω behaves like a strong inaccessible. It is standard simply to adopt the existence of ω as a further axiom. From a purely iterative point of view, this has always seemed a great leap, as indeed it is. And, as Russell stressed, any assumption of infinitely many urelements, however favoured by one or another physical theory, should form no essential part of pure mathematics. On the other hand, there is the Fregean approach, via comprehension: having identified the natural numbers somehow as mathematical objects (e.g. sets à la von Neumann or Zermelo), the extension of 'is a natural number' will be an infinite set, as desired. But, as usually presented, such an

---

[27] Of course, if we work within ZF, i.e. we adopt the ZF axioms categorically and describe full models as sets, then this cannot happen. The problem arises when we shift to the structuralist viewpoint. Then we only have the modal existence of full models and the EP as categorical assumptions (beyond the background logic). We have then only assumed Replacement within a possible model, not as a principle governing relations among many possible models. $\vDash_P$ can still be introduced inductively and stability can be proved. But, as the example shows, verification of the ZF axioms in the sense of $\vDash_P$ is not automatic.

argument would use unrestricted comprehension (for ∈-formulas), but then we prove too much. Zermelo's remedy, through *Aussonderung*, restricts comprehension to subsets of a given set, and then, of course, the argument is circular. So the Axiom of Infinity is simply taken as basic.

It will be recalled from Chapter 1, however, that the modal-structuralist framework we have been employing throughout has the resources for restoring a Fregean derivation of Infinity, without paradox and without violating Russell's requirement of "no contingent infinities".

For convenience, let us repeat the core of that argument here. One begins with a constructive rule for generating a "next object", say a stroke, from any finite sequence of given objects of the same sort. Such objects are assumed to be first order. Call the rule R, and let $P(x, y)$ be the predicate (of "strokes"), "y is generated after x in accordance with R". Now the range of P, $Rn(P) = \{y \mid \exists x(P(x, y))\}$, may in fact be finite, but it could, logically, have been infinite. More precisely, we can adopt the following as an axiom:

$$\Diamond[\exists x \exists y(P(x, y)) \,\&\, \text{"P is asymmetric and transitive"}$$

$$\&\ \forall x \exists y(P(x, y))], \tag{*}$$

where the quoted portion abbreviates the familiar first-order conditions. The next step is to apply necessitated comprehension of second-order logic:

$$\Box \exists X \forall z(X(z) \equiv \exists x(P(x, z))),$$

guaranteeing that "in any world" P has a range. But, of course, it can also be proved that

$$\Box[(*)^- \supset \text{"}\exists X(X \text{ is } Rn(P) \,\&\, X \text{ is infinite"})],$$

where $(*)^-$ is (*) without the initial $\Diamond$, and the quoted portion is spelled out in second-order notation (employing either "X includes an ω-sequence" or "X is Dedekind-infinite"). But then from (*) and modal logic we have

$$\Diamond \exists X(X \text{ is infinite}),$$

as desired. It is only $\Diamond \exists$ that matters, so that no contingent actual infinities need be assumed. Second-order comprehension treats the predicate, "is generated in accordance with rule R", as it does any other predicate of first-order objects. It has an extension. (No inconsistency here!) Logically, it could have an infinite extension, and that is all that mathematics requires of its first axiom of infinity.

As was noted in Chapter 1, the assumption (*) is grounded in a constructive process and is of the sort that even constructivist opponents of classical reasoning about the infinite have been prepared to grant (cf. especially Dummett [1977]). However, it must be conceded that (*) is not as innocent as it might be from a constructivist standpoint. The really innocent reading of the statement, "Any finite sequence of strokes can be extended", could be written,

$$\Box \forall x \Diamond \exists y (y \text{ extends } x)$$

(where we understand the quantifiers as restricted to objects generated in accordance with a rule such as R). Here we contemplate moving from any world in which a finite sequence has been generated to a (possibly different) world in which the sequence is extended. No single world need be recognized satisfying

$$\forall x \exists y (y \text{ extends } x).$$

But such a world is precisely what is contemplated in (*). A strict finitist cannot be forced to accept it. It could dogmatically be insisted, for example, that standard Euclidean models of space and time are logically impossible! Thus, even with modality, the first axiom of infinity cannot be made innocuous.

On the possible worlds metaphor, the axiom postulates a fixed point of an operation on worlds: the operation involves looking at the finite segments, σ, generated in a world w according to a (fixed) constructive rule, and moving to a w' (uniquely selected somehow) in which each of those segments has a proper extension. A fixed point of this operation is a world w^* such that $w^{*'} = w^*$; then w^* answers to (*) and must be infinite. It is interesting to observe that stronger set-theoretic principles, functioning as axioms of infinity, can be obtained in a similar fashion.

Consider Replacement (the second-order version, the one we naturally read and assent to when learning the ZF axioms). So far we have taken it for granted, in the sense of postulating the possibility of a ZF^2 model. I would suggest that one reason why this seems reasonable is due to an analogy with the first axiom of infinity, as just discussed. The analogy depends on reading and accepting the

Extendability Principle (EP) as applying to models of Zermelo set theory (Z) (= ZF less Replacement but with *Aussonderung*), i.e.

$$\Box \forall M \Diamond \exists N (N \text{ properly extends } M),^{28} \qquad \text{(ZEP)}$$

where the quantifiers are understood as relativized to Z models, spelled out in second-order form in the by now familiar way. Now consider any possible Z model M and any function $F: M \to M$. For any "set" a of M, $F[a]$ need not be a "set" of M, but it will at least be a sub-collection of the domain of M. But when we move to any proper extension N of M, N will contain $F[a]$ as a "set", since the power set operation will have been applied to the domain of M in N. Consider the operation of moving from M to the smallest extension N such that N contains as a "set" the range of every $F: M \to M$ on every a in M. We obtain (the possibility of) Replacement by postulating a fixed point of this operation. For such a fixed point will be an M^* such that it already contains as a "set" $F[a]$ for arbitrary "sets" a of M^* and $F: M^* \to M^*$.[29]

To be sure, this is no derivation of Replacement. It is at best a replacement of a derivation: a placement of Replacement in a new context, one which brings out an analogy with the first axiom of infinity: both have a similar motivation in extendability principles, which automatically guarantee a requirement of richness with respect to objects of any given model. The extra strength comes in postulating a fixed point in the relevant process of taking extensions. As we shall see, further leaps may be motivated along similar lines.[30]

---

[28] In order to guarantee that a proper extension N of M effectively contain $|M|$ as an element, the requirement that N contain an equivalent, in the sense of n. 22, should be added as part of the definition of "proper extension". Lacking Replacement, we can no longer derive this condition from the ordinary weaker notion of proper extension, as we could in the case of ZF models.

[29] Strictly speaking, such talk of operations on models is naturally formalized as third order (if we insist on formalizing it). But it can be reduced to second order by treating models operated on as (first-order) elements of any hypothetical background model (which we have already done, anyway, in speaking of proper extensions).

[30] It is interesting to compare the motivation just given for Infinity and Replacement with Parsons' derivation of the modal translations of these axioms, in "theories of potential sets", from a principle allowing passage from certain statements of the form $\Box \forall x \Diamond \exists y \, A(xy)$ to those of the form $\Diamond \forall x \exists y \, A(xy)$ (in Parsons [1983], p. 323). This was precisely the move we found the strict finitist capable of resisting, above, in connection with the Dedekind-inspired derivation of Infinity. Mathematically, the derivation given by Parsons corresponds to the well-known derivation of Infinity and Replacement from the Reflection Principle. (I am indebted to Charles Parsons for calling this matter to my attention.) On the present approach, postulation of fixed

§ 5. Axioms of Infinity: Climbing up

Having motivated Replacement[2], the next step is to adopt it as a structure-characterizing axiom. In the present modal-structuralist framework, this amounts to adopting (2.4) above, asserting the possibility of a ZF^2 model (stated in purely second-order terms). Such an assertion of possibility replaces the standard platonist "adoption of axioms" as truths about a fixed set-theoretic universe.

Once Replacement[2] has been adopted in the modal-structuralist sense, the Extendability Principle (EP) applied to ZF^2 models—i.e. (2.5) above—yields strongly inaccessible cardinals, as already observed. The height of any possible full ZF^2 model, M, is such a cardinal and occurs in any proper extension of M. And since the EP implies that the possibilities of extensions go on and on, one obtains the Putnam-translate of the Axiom of Inaccessibles, i.e. of

$$\forall\alpha\exists\beta(\beta > \alpha \ \& \ \mathrm{Inac}(\beta)). \tag{AI}$$

The next step (upward) is to adopt *this* as a structure-characterizing axiom, i.e. to adopt

$$\Diamond\exists X\exists f(\wedge ZF^2 \ \& \ AI)^X\!\binom{\in}{f}, \tag{2.6}$$

the intrinsic modal second-order assertion of the possibility of a full model of ZF^2 plus the Axiom of Inaccessibles. And, together with this, one naturally rewrites the EP so as to assert the possibility of properly extending any model of ZF^2 & AI. Let M be any such model and N a proper extension. Then in N, the height of M exists and is a hyperinaccessible cardinal, that is, a solution to $\phi_\kappa = \kappa$, where ϕ_α is an enumeration, in order, of inaccessibles (α an ordinal of N). (κ is strongly inaccessible with κ strongly inaccessible below it.) Thus, the new EP guarantees endlessly many hyperinaccessibles, hence the Putnam-translate of,

$$\forall\alpha\exists\beta(\beta > \alpha \ \& \ \beta \text{ is hyperinaccessible}), \tag{AH^1I}$$

the Axiom of Hyperinaccessibles. If one then adopts *this* as a structure-characterizing axiom (on the plan of (2.6)) and modifies the EP accordingly, one obtains hyper-hyperinaccessibles, and (the

points of certain extension-taking operations (an approach "from below"), replaces reflection principles (which proceed "from above"). For further discussion, see below, Ch. 2, § 5.

Putnam-translate of) the Axiom of Hyper-hyperinaccessibles (AH^2I), and so on.

It is worth noting that, at each stage in this process, the adoption of the new structure-characterizing axiom can also be seen as the postulation of a fixed point of a suitable operation on models, in analogy with the procedures of the previous section corresponding to Infinity and Replacement. Consider, for example, the step from inaccessibles to hyperinaccessibles. Let M be any full ZF^2 model and let $O(M)$ be the smallest $N \geq M$ such that, for each ordinal α of M, N contains an inaccessible κ such that $\kappa > \alpha$. In general $O(M) \gneqq M$ (and κ can be found independent of α), but in case $O(M) = M$, M then satisfies the axiom of inaccessibles and so has hyperinaccessible height. Thus, postulating a fixed-point model for O is equivalent to adopting the new axiom, in this case the Axiom of Inaccessibles.

A next major step (further upward) is to the so-called Mahlo cardinals. It is known how these may be motivated by reflection principles.[31] Alternative motivation can be given, "from below" as it were, along lines similar to the above, employing suitable extendability principles together with appropriate structure-characterizing axioms, which in turn can be obtained by postulating fixed points on certain extension-taking operations on models. Let us see how this works in the case of strongly Mahlo cardinals.

A cardinal κ is *strongly Mahlo* just in case every normal function f on κ has a strongly inaccessible fixed point, where a *normal* function f on κ is a function from ordinals to ordinals $< \kappa$ which has range unbounded in κ, is increasing, and is continuous at limits (i.e. for limits $\lambda, f(\lambda) = \bigcup_{\xi < \lambda} f(\xi)$). If one adds to ZF^2 the axiom,

$$\text{"Every normal function has an inaccessible fixed point"}, \qquad \text{(F)}$$

one obtains a theory whose full models M are of the form V_κ with κ (strongly) Mahlo.[32] This can all be stated in the intrinsic second-order form we have been using throughout, and the proof that the height of any model, M, of $ZF^2 + F^2$ is Mahlo can be carried out

---

[31] See Lévy [1960]; cf. Drake [1974], Ch. 4, §4, and below.

[32] NB. This depends critically on taking (F) as a second-order axiom, quantifying over arbitrary functions from ordinals to ordinals. If, instead, the first-order instances of the scheme are taken as axioms (i.e. one for each formula $\psi(x, y)$ defining a normal function), the height of a model V_κ of ZF + the F instances need not be Mahlo. The situation parallels that already encountered in connection with Replacement and inaccessibles. Cf. Drake [1974], Ch. 4, Ex. 3.7(3). Incidentally, the label 'F' is Drake's.

relative to any proper extension of M. (Here and below, 'F^2' denotes the second-order statement of (F).) Thus, once we have adopted Axiom F^2, the EP applied to models of ZF^2 + F^2 guarantees (endlessly many) Mahlo cardinals, as "elements" of extensions. (In brief: Axiom F^2 is to Mahlo cardinals what Replacement^2 is to inaccessibles.)

What can be said by way of motivation for adopting Axiom F^2 (i.e. in the sense of the possibility of a full model of ZF^2 + F^2)? At least a strong analogy with Replacement^2 can be seen on the present approach. Given a full model M of ZF^2, one can consider the operation O of passing to the least extension $N \geq M$ such that, for any normal function f: $\mathrm{Ord}(M) \to \mathrm{Ord}(M)$ (i.e. from the ordinals of M to the ordinals of M), N contains an inaccessible fixed point either of f or of any normal proper extension g of f. Now observe that, for any full model M of ZF^2 and any normal f: $\mathrm{Ord}(M) \to \mathrm{Ord}(M)$, the height of κ of M will be an inaccessible fixed point of any normal g properly extending f. (κ is inaccessible and $g(\kappa) = \bigcup_{\xi < \kappa} g(\xi) = \bigcup_{\xi < \kappa} f(\xi) = \kappa$, the first equation by continuity.) Thus any proper extension of M will satisfy the condition, so that if $O(M) \neq M$, $O(M)$ will be the least proper extension of M. Now consider a fixed point model M^* of this operation O, i.e. $O(M^*) = M^*$. Then for any normal f: $\mathrm{Ord}(M^*) \to \mathrm{Ord}(M^*)$, M^* already contains an inaccessible λ such that $f(\lambda) = \lambda$, i.e. an inaccessible fixed point of f. Thus, M^* satisfies Axiom F^2.

Thus, just as in the case of Replacement^2, proper extensions are automatically rich enough to fulfil the desired condition with respect to all functions on the given model: the desired axiom, in the relevant form of modal existence of a full model, says that there could be a model (of the weaker theory, ZF^2 in the case of F^2) already sufficiently rich in the relevant respect. The same sorts of steps can be repeated to give higher and higher types of Mahlo cardinals.

The game of large cardinals has been played to staggering heights, and so far we have only motivated some of the very tiniest that have been defined and explored. It is noteworthy that nearly all of these can be arranged on a linear increasing scale, although the methods for their generation are remarkably diverse.[33] A theoretically important break is considered to occur with partition cardinals,

[33] For a breath-taking survey, see Kanamori and Magidor [1978]. See also Maddy [forthcoming] for valuable information on informal motivating principles behind various large cardinal axioms.

beyond which the structurally interesting Axiom of Constructibility ("V = L") is known to fail.[34] Such cardinals are really large (sometimes called "large, large", as contrasted with the "small, large" ones we have been dealing with), and would appear to be beyond the reach of anything like the modest methods employed here. But—in the spirit of true climbing—the name of our game is not height for height's sake, but rather to explore the potential of certain quasi-constructive methods closely connected to fundamental iterative principles. How far do they extend?

Without pretending to give a final or precise answer, we can observe that difficulties arise already at the level of the so-called "weakly compact" cardinals, the next big step up beyond the Mahlo cardinals (as the route is usually described). (The weakly compact cardinals are definable in several very different sounding ways, one of which will be utilized momentarily. They are still "small" relative to the theoretically important break just cited, i.e. they are compatible with V = L.)

The notion of weak compactness most readily adaptable to the present methods is that involving trees:

> A cardinal κ has the *tree property* iff every tree T ($= \langle T, <_T \rangle$) of cardinality κ such that every level of T has cardinality $< \kappa$ has a branch of cardinality κ.

A cardinal κ is *weakly compact* iff it is strongly inaccessible and has the tree property.[35] That ω has the tree property is the content of the König infinity lemma; weak compactness is a direct generalization of this.

Now, to obtain weakly compact cardinals on the present approach, one must formulate a sentence, S, of second-order logic (in effect, a Π^1 or Σ^1 sentence) such that it can be proved that any full model of $ZF^2 + S$ has weakly compact height. Postulating the possibility of such models and applying the EP will then give (indefinitely many) weakly compact cardinals, i.e. the Putnam-translate of

$$\forall \alpha \exists \beta (\beta > \alpha \ \& \ \beta \text{ weakly compact}).$$

---

[34] Key results in this area are due to Scott, Silver, and others. Details and references can be found in Jech [1978] and in Drake [1974].

[35] Sometimes the requirement of strong inaccessibility is omitted from the definition and tacked on later when equivalence with other definitions is proved.

Next one would like to be able to motivate adoption of the modal-existence postulate (asserting the possibility of a full model of ZF^2 + S) in the manner invoked above, that is, by seeing the postulate as a fixed point of an operation of taking extensions.

Now the first task is readily accomplished. One can formulate a second-order tree axiom with the desired properties as follows:

$\forall T \forall R[\{\langle T, R \rangle$ a tree & $\forall \alpha(\alpha$ an ordinal $\supset$
T has an element of level α) &
"Every level of T is a set"$\} \supset \exists B\{B$ a branch of T &
$\forall \alpha(\alpha$ an ordinal $\supset$ ^{B} has an element of level $\alpha)\}$]. (TA)

(Here the capitalized quantifiers are second order; the quantifiers over levels are second order, and (after employing the usual definition of 'level') "level L is a set" can be expressed simply as '$\exists x \forall y(L(y) \equiv y \in x)$', guaranteeing that L has a cardinality, which is what matters here.) Now one can prove that any full model of ZF^2 + TA is of the form V_κ with κ weakly compact. (This can be given in the relevant intrinsic second-order form as in the cases of inaccessibles and Mahlo cardinals, noted above.)

The problem, however, concerns the next step, motivating the modal-existence postulate. How can the possibility of a full model of ZF^2 + TA be made plausible on the basis of taking extensions of ZF^2 models? In the cases of Replacement and Axiom F, extensions of any model of the theory less the respective axiom fulfilled the demands of that axiom with respect to the objects of the given model. But TA has a different structure; it says that counter-examples to the generalization of the König infinity lemma are not possible, and one cannot automatically remove such a counter-example merely by moving to a proper extension.

This is not to say that there is no good motivation for weakly compact cardinals, but simply that such motivation must come from elsewhere. One source is by a simple analogy with ω, which has the tree property. (Shouldn't an uncountable inaccessible with the tree property also be a possibility?) Perhaps somewhat more compelling is an analogy with inaccessibles, via notions of indescribability.[36] Given a class, Ω, of formulas with only second-order variables $X_1, \ldots, X_n$ free,

---

[36] For a survey of results, see Drake [1974], Ch. 9, and Jech [1978], pp. 385 ff.

an ordinal α is Ω-*indescribable* iff no formula ϕ in Ω describes α, where this means that for no ϕ in Ω is it the case that

$$\langle V_\alpha, \in | V_\alpha \rangle \vDash \phi(U_1, \ldots, U_n), \text{ for some } U_1, \ldots, U_n \subseteq V_\alpha,$$

but for all $\beta < \alpha$,

$$\langle V_\beta, \in | V_\beta \rangle \nvDash \phi(U_1 \cap V_\beta, \ldots, U_n \cap V_\beta).$$

(In case a formula ϕ meets this latter condition, α is said to be *described by* ϕ.) Now it can be proved that the strongly inaccessibles are just the first-order indescribable cardinals (i.e. the Π^1_0-indescribables).[37] And it also can be shown that the weakly compact cardinals and the Π^1_1-indescribable cardinals coincide.[38] If one accepts inaccessibles—i.e. Π^1_0-indescribables—should one not also accept weakly compact cardinals? And other analogies to smaller large cardinals have been suggested, based on considerations of inductive definition (cf. Tait [forthcoming]). But however strong these motivations may be, it seems interesting that the sorts of appeals to extendability principles we have made do not carry us this far.

Still, the importance of motivating the smallest large cardinals—however minuscule they can be made to appear—should not be underestimated. They are, of course, central to the whole subject of large cardinals, and whatever reasons can be given for accepting them (i.e. their possibility) can affect our view of that subject. Usually one approaches large cardinals only after learning the ZF axioms, and—in the face of their indemonstrability in ZF—one thinks of them as esoteric. On the view we have been exploring, this is a misleading picture. The extendability of ZF structures is firmly rooted in our informal notions of iteration, perhaps as firmly as the very notions formalized in ZF itself. Stopping with Transfinite Induction (or Replacement) seems to us arbitrary. From this perspective, the study of large cardinals is an intrinsic and virtually inevitable part of set theory.

In closing, it is worth making a brief comparison with alternative ways of motivating inaccessible and Mahlo cardinals. Perhaps the most standard is by appeal to "reflection principles". These come in various forms; some are expressible in the language of ZF, others re-

---

[37] It should be noted that this depends on the use of second-order parameters in the definition of indescribable. If they are omitted, Replacement will allow accessibles to be first-order indescribable (cf. Drake [1974], pp. 268–76, esp. Ex. 1.11).

[38] For the proof, see e.g. Drake [1974], pp. 292, Theorem 2.1.

quire the language of proper classes and beyond (cf., e.g., Reinhardt [1974*a*, *b*]). But all have in common the idea that what holds in the Cantorian universe, V, ought already to hold "lower down", i.e. in a restricted initial segment of V. In fact, in so far as first-order properties of sets are concerned, an important formalization of this is provable in ZF:

$$\text{ZF} \vdash \forall\alpha\exists\beta > \alpha\forall x_1 \dots \forall x_n \in V_\beta[\phi(x_1,\dots,x_n) \equiv \phi^{V_\beta}(x_1,\dots,x_n)],$$
$$(R_0)$$

where ϕ is any formula of ZF with just the x_i free (and lacking abstraction terms), V_β is the set of sets of rank $< \beta$ (i.e. the cumulative hierarchy up to β), and ϕ^{V_β} is the relativization of ϕ to V_β. R_0, it turns out, is equivalent to the combination of the axioms of Infinity and Replacement (the first-order scheme). Of course, it does not yield large cardinals, but strengthenings of it do. For example, if to R_0 the condition is added that β be inaccessible, the result, R_1, can be proved in ZF to be equivalent to Axiom F (cf. Drake [1974], p. 121), and therefore implies the existence of many kinds of inaccessibles. This is useless as motivation for simply strong inaccessibles, of course, since the Axiom of Inaccessibles (AI) has been built into R_1. But once given R_1 hyperinaccessibles, etc. follow. And, if we permit ourselves talk of V, Axiom F leads us to describe the proper class, Ω, of all ordinals (i.e. all ordinals $\in$ V) as a Mahlo cardinal. (This depends on reading Axiom F not as a scheme—whose instances R_1 implies—but as the second-order statement F^2.) If one then (informally) applies reflection to *this* property of Ω (or to the related property of V), one concludes that some cardinal κ (i.e. a $\kappa < \Omega$, hence a set) must be Mahlo as well. Further applications of Reflection can be used to argue that in fact there must be arbitrarily large Mahlo's, and one can then adopt R_2, just like R_1 except for containing the condition that β be Mahlo instead of merely inaccessible. This in turn will be equivalent to a scheme asserting that every normal function has a Mahlo fixed point; the second-order version of this makes Ω hyper-Mahlo, etc., etc.,…, etc.,…

In fact, direct reflection on the universe as a mathematical object has been proposed as a key to motivating much larger cardinals (e.g. measurable cardinals, incompatible with V = L).[39] And, of immediate concern to us here, it has been proposed as the natural way to

---

[39] See Reinhardt [1974*b*], which employs systems axiomatizing properties of proper classes.

get strongly inaccessibles in the first place.[40] Obviously, Ω, if we recognize it, must be strongly inaccessible. (The cardinality of the power set of any set is a set, and the limit of any sequence of ordinals ($<\Omega$) of length $<\Omega$ is a set.) So, reflecting on *this* property of Ω, we conclude that there must be a set which is a strongly inaccessible cardinal.

To give one further illustration of such reasoning, let us use it to argue directly for Axiom F^2, the full second-order version. Let f be any normal function, $f: \Omega \to \Omega$. Let g extend f by taking just Ω as argument. If g is normal, $g(\Omega) = \Omega$, so that g has an inaccessible fixed point. (Or, we can simply stipulate that $g(\Omega) = \Omega$, as far as this argument is concerned.) By reflection on *this* property of Ω, g, hence f, must have an inaccessible fixed point lower down, i.e. at some set cardinal. But f was arbitrary, so F^2 holds. Further reflection yields Mahlo cardinals as already observed.

There are two main objections to this sort of reasoning "from above". The first concerns reflection itself. Even if proper classes $(V, \Omega,$ etc.) are admitted as objects, what properties of these are reflectable? Obviously, not arbitrary properties, e.g. "containing all set-ordinals". In the case of R_0, the provable Reflection Principle, no distinction need be drawn, but in the above higher-order arguments, such a distinction is crucial. Why should 'being inaccessible' or 'being a fixed point of an extension of a normal function' be reflectable? Whatever answer we try to give, it seems that we end up appealing to the possibility of sufficiently rich models, which is exactly what the present approach from below involves. Second, from the present point of view, of course, the appeal to proper classes as objects is a cardinal sin (inaccessible or not!). Not only does it involve treating as "completed" what is not supposed to be completable; if one is prepared to speak of functions taking Ω as argument or value, one is prepared to perform set-like operations on proper classes after all, and it then appears that V, Ω, etc., are simply behaving as an inaccessible level of sets.

In contrast, the above-sketched approach "from below", while frankly admitting the need for new axioms along the way (modal existence of fixed-point models, and extendability principles), achieves the goal of generating small large cardinals on the basis of a steadfast rejection of proper classes as objects. Dealing with many

---

[40] See Reinhardt [1974a].

possibilities of set-theoretic structures (in the spirit of Zermelo and Putnam) is no doubt a bit more complex than working with a single fixed universe, but it has its rewards—and it may ultimately help in the effort to sustain Cantorian insights.[41]

[41] For a valuable survey of Cantorian thought, see Hallett [1984]. In further work, I hope to explore some of the implications of the foregoing for Hallett's critique of limitation of size arguments.

Appendix

Here we give proofs of the stability and correctness theorems for Putnam semantics stated in § 3 of the text. Recall that Putnam semantics, the theory of $\vDash_P$, was introduced in § 3 as a relation between full ZF^2 models and sentences of the language of ZF^2. In consequence of the Zermelo quasi-categoricity theorems, we may restrict attention to models linearly ordered by the relation of end-extension, i.e. for any pair, M, N, of such models, either $M \leq_e N$ or $N \leq_e M$. The stability of $\vDash_P$ is given by the

Stability Theorem: Let A be a sentence of $\mathscr{L}(ZF^2)$ and M a full ZF^2 model; then

$$M \vDash_P A \text{ iff } \forall M' \geq M \; M' \vDash_P A.$$

This follows from a more general result for formulas (allowing for free variables):

Lemma: Let $A(u_1 \ldots u_n)$ be any formula of $\mathscr{L}(ZF^2)$ and M a full ZF^2 model and E an evaluation of variables over M. Then

$$M \vDash_P A(u_1 \ldots u_n) [E] \text{ iff } \forall M' \geq M \; \forall E', \text{ an evaluation of variables over}$$
M' agreeing with E on $u_1 \ldots u_n$, $M' \vDash_P A(u_1 \ldots u_n) [E']$.

Proof by induction on the number of quantifiers in A. (We write this for the case of first-order formulas for ease of exposition. Modifications for second-order formulas are straightforward.)

(i) A is quantifier free. Then the result follows by the clauses 1–3 in the definition of $\vDash_P$ and standard facts about ordinary $\vDash$.

(ii) A is $\exists x B$. (Again, for ease of exposition, let us revert to ordinary substitution notation in place of quantification over evaluations of variables.) Let $M \vDash_P \exists x B$ (i.e. at any E). Then $\exists N \geq M \; \exists z \in |N|$ such that $N \vDash_P B(x/z)$, whence $N \vDash_P \exists x B$. Now let $M' \geq M$. By the linear ordering of relevant models, there are two cases to consider:

(a) $M' \geq N$. Then by inductive hypothesis, as $N \vDash_P B(x/z)$, $M' \vDash_P B(x/z)$, whence $M' \vDash_P \exists x B$ (by clause 4 of $\vDash_P$).

(b) $N \geq M'$. Then, as $N \vDash_P B(x/z)$, by clause 4 of $\vDash_P$, $M' \vDash_P \exists x B$.

(iii) A is $\forall x B$. Assume $M \vDash_P \forall x B$. Then, by definition,

$$\forall N \geq M \; \forall z \in |N| \; N \vDash_P B(x/z). \tag{*}$$

Let $M' \geq M$. Then, by transitivity of $\geq$ and (*),

$$\forall M'' \geq M' \; \forall z \in |M''| \; M'' \vDash_P B(x/z).$$

Therefore $M' \vDash_P \forall x B$, by clause 4 of $\vDash_P$. This completes the proof.

The correctness of $\vDash_P$ asserts that sufficiently rich ZF^2 models treat first-order sentences on $\vDash_P$ just as the ZF universe V treats those sentences on ordinary $\vDash$. This can be proved in NBG together with the Axiom of Inaccessibles.

Correctness Theorem: Let $A(a_1 \ldots a_n)$ be any ZF^1 sentence with the parameters displayed (i.e. the a_i are sets of V assigned to free variables $x_1 \ldots x_n$ by a given evaluation of variables), and let $V \vDash A(a_1 \ldots a_n)$. Then there is a κ such that V_κ is a full ZF^2 model, the $a_i \in V_\kappa$, and $V_\kappa \vDash_P A(a_1 \ldots a_n)$.

Proof by induction on the number of quantifiers in A.

(i) A is quantifier free: Then for κ inaccessible and $> \max(\rho(a_i))$ (where $\rho(x)$ is the ordinal rank of x), $V_\kappa \vDash_P A$, by clauses 1–3 of the definition of $\vDash_P$ and the fact that V_κ is a transitive submodel of V.

(ii) A is $\exists x B(x, a_1, \ldots, a_n)$: By assumption, there is a set c such that $V \vDash B(c, a_1, \ldots, a_n)$. By inductive hypothesis, $\exists \kappa > \max(\rho(c), \rho(a_i))$ such that $V_\kappa \vDash_P B(c, a_1, \ldots, a_n)$, whence $V_\kappa \vDash_P \exists x B(x, a_1, \ldots, a_n)$.

(iii) A is $\forall x B(x, a_1, \ldots, a_n)$: Let κ be the least inaccessible $> \max(\rho(a_i))$. We show that $V_\kappa \vDash_P A$. For this, we need to show that

$$\forall \kappa' \text{ such that } V_{\kappa'} \vDash ZF^2 \text{ and } V_{\kappa'} \geq V_\kappa, \forall c \in V_{\kappa'}:$$

$$V_{\kappa'} \vDash_P B(c, a_1, \ldots, a_n). \tag{*}$$

Let κ' and c be as in the hypothesis of (*). By hypothesis of the theorem, $V \vDash B(c, a_1, \ldots, a_n)$. By inductive hypothesis, there is inaccessible κ'' such that $V_{\kappa''} \vDash_P B(c, a_1, \ldots, a_n)$ (with $c, a_i \in V_{\kappa''}$). In analogy with the corollary to the Stability Theorem (Chapter 2, § 3), it follows from the lemma for the Stability Theorem above that $V_{\kappa'} \vDash_P B(c, a_1, \ldots, a_n)$ also. But κ' and c are arbitrary, so (*) holds, whence $V_\kappa \vDash_P \forall x B(x, a_1, \ldots, a_n)$. This completes the proof.

3

Mathematics and Physical Reality

§0. Introduction

As is universally recognized, mathematics throughout its history has been intimately bound up with our interactions with the material world, from the most mundane practical enterprises of counting and measuring to our most sophisticated theoretical efforts to comprehend its workings as the unfolding of physical laws. From a historical perspective, it would be no exaggeration to say that physical application has sustained mathematics as its very life-blood.

This perspective is reflected in philosophy of mathematics. Surely one of the strongest reasons—if perhaps not the only reason—for taking mathematical truth seriously stems from the apparently indispensable role mathematical theories play in the very formulation of scientific descriptions of the material world around us. As soon as we undertake to convey the information that, say, there are more spiders than apes, we seem to be committed to numbers, or classes and functions. Describing the behaviour of the stars and galaxies apparently involves us in a good deal of the apparatus of differential geometry. And to probe the atomic structure of matter and the underpinnings of chemistry and biology, we seem to be involved in the theory of Hilbert space and a generalized form of measure theory. Whatever the details of this entanglement between physics and mathematics, surely a purely formalist approach to mathematics would seem far more plausible were there no entanglement. If we strain our imaginations and suppose (*per impossibile*!) that mathematical theories and structures had no material applications—that they could somehow be isolated from the empirical sciences—what objection would we have to treating mathematics as a purely formal game? For such a "mathematics", the question of truth might not even seem to arise.

Reflections such as these lead us to pose three interrelated fundamental questions concerning mathematics in its applications. The first is this: granted that the role of mathematics in ordinary and scientific applications provides some grounds for taking mathematical

truth seriously (that is, for taking a realist as opposed to an instrumentalist stance toward at least some mathematical theories), are
these the exclusive grounds, or are there others; and, if there are
others, what are they and how powerful are they? The second
question—really a composite of questions—asks how much mathematics is really indispensable for how much science? And, third, we
must ask, just what does such indispensability demonstrate with
regard to mathematical objectivity and mathematical objects?

Having now perhaps piqued the reader's interest in the general
subject, I must offer a disappointing apology in advance: none of
these questions will be answered definitively. At best, partial and
tentative answers to the second and third questions will emerge.
As to the first, we shall be left hanging.

For the task that demands immediate attention is that of sketch-
ing how the modal-structuralist framework already developed for
pure mathematics can be extended to applied mathematics. How, in
the first place, are we to represent ordinary and scientific applied
mathematical statements? What are the main assumptions that lie
behind such a representation? Having sketched the basic ideas and
broached some of the main problems (in §1), we may then turn our
attention back to the broader questions concerning indispensability.
As we shall see (in §2), there is a strong case that modern physical
theories—especially General Relativity and Quantum Mechanics—
require (the possibility of) mathematical structures so rich that even
the chances of a "modal nominalism" in any reasonable sense are
dim. This case, as we shall present it, depends on a rather broad
interpretation of 'applied' in the phrase 'applied mathematics': questions of theoretical physics of a foundational character are included.
But we see no rational basis for excluding them (e.g. by drawing a
sharp line between "ordinary empirical" applications and "theoretical" or even "meta-theoretical" applications). And, it would be
ironic indeed if foundations of mathematics took the stance that
foundations of physics need not be respected!

In fact, we shall further see that recent work on the implications of
higher set theory raises the tantalizing prospect that stronger and
stronger abstract mathematical principles may have consequences of
physical significance, undecidable in weaker theories, suggesting that
it would be futile to seek any a priori global framework for applied
mathematics. (This will be brought out in §3.)

As we have already seen, the modal-structural treatment of pure mathematics invokes counterfactuals of a strict kind: all relevant conditions can be stated in the antecedents (as the categoricity proofs show, where applicable). Thus, the notorious problems concerning what "relevant background conditions" are to be understood as held fixed in interpreting ordinary counterfactuals (associated with the "problem of cotenability", cf. above, Chapter 2, § 1) did not arise in the context of pure mathematics. When we turn to applied mathematics, however, there is a sense in which the problem returns to haunt us, as we shall soon see.

In the final sections (§§ 4 and 5), various approaches to this problem will be explored. How one reacts to it, in fact, seems to depend on one's "realist commitments" concerning non-mathematical reality. If one's "realism" is sufficiently strong, the problems seems to evaporate. But if the ms approach seeks to maintain a "metaphysical neutrality" on such questions, thorny problems arise in the very formulation of the applied mathematical counterfactuals. Efforts to overcome them raise some interesting points of comparison with recent "synthetic" approaches to physical theories (motivated by nominalist concerns and aimed at challenging the alleged scientific indispensability of mathematics entirely).[1] As we shall see, some of the technical portions of such work (e.g. Field-style representation theorems) can be of relevance to a ms programme, but such theorems go beyond what is required in crucial ways. And, from our own perspective, the phenomenon of non-conservativeness of mathematically rich theories (highlighted in § 3) tends to undermine any sweeping challenge to the indispensability of "abstract" mathematical theories.

Of necessity, we have concentrated here on problems of formulation involved in a ms treatment of applied mathematics, and on some of the technical and philosophical questions immediately surrounding these problems. However, the broader questions of justification of the ineliminable postulates of ms mathematics—especially the modal-existence axioms—must not be forgotten. In this connection, applied mathematics can provide a crucial epistemological link, much as it has been thought to provide under familiar platonist treatments. The point here is to adapt Quinean indispensability arguments to the modal framework: rather than commitment to certain abstract objects receiving justification via their role in scientific

---

[1] See Field [1980].

practice, it is the claims of possibility of certain types of structures that are so justified. Moreover, to the extent that indispensability arguments can be adapted to the modal approach, their usual platonist thrust is actually undermined: a fixed realm of abstract objects is not really shown to be indispensable; rather it is the weaker claims of possibility that occupy such a position.

§1. The Leading Ideas

Much as in the case of pure mathematics, we may attempt to represent ordinary applied mathematical statements as elliptical for modal conditionals of a specified form. Such conditionals spell out what would obtain were there any suitably rich (pure) mathematical structure in addition to the actual non-mathematical objects or systems to which we are applying mathematical concepts and theories. Here the modality of the counterfactual is a logico-mathematical one, just as in the treatment of pure mathematics. Although we may be applying mathematics to physical objects, we are not automatically constrained to hold physical or natural laws fixed in contemplating a purely mathematical structure in addition for the purposes of carrying applied mathematical information. (Thus, for example, we are free to entertain the possibility of additional objects—even physical objects—of a given type, to serve as components of a mathematical structure. Such objects could be conceived as occupying a certain region of space-time but as not subject to certain dynamical laws normally stated universally for objects of that type.) Just what must be held fixed is a matter to which we shall return below.

But how, it may be asked, can an additional structure for pure mathematics (such as an ω-sequence or a complete ordered separable continuum) be brought to bear on material objects? Imagining such a structure—whether thought to occupy space-time or not— does no good unless we can also speak of relations between the material system in question and items of the mathematical structure. Thus, to represent simple counting, for example, it does little good to entertain the possibility of an ω-sequence in addition to the actual objects to be counted unless we can also speak of correspondences between those objects and the items serving as "numbers" of the hypothetical ω-sequence.

One solution to this problem is to move immediately to models of set theory, that is, to entertain hypothetically models of a suitable set theory in which actual objects are taken as urelements. Then we have the operation of set formation applied to those actual objects, and the usual apparatus of mappings and number systems (set-theoretically construed) is available. This might be called the *global approach*, since, if the set theory chosen is sufficiently rich, it can be invoked to handle virtually any present or foreseeable instance of applied mathematics.

While there is a good deal to be said in favour of such an approach (especially concerning its intuitiveness, its power, and its simplicity), there is also independent interest in pursuing a *piecemeal approach*, in which we limit the hypothetically entertained (pure) mathematical structures to a level that is actually needed for the purposes of the application in question. In part, such an approach is motivated by an independent interest in the second leading question posed above ("how much mathematics for how much science"), which the piecemeal approach is forced to face. There are also legitimate concerns over the consistency of powerful set theories, and over their "abstractness". Do we really need to iterate the power set operation beyond anything that we could be said to experience, beyond say the level of space-time regions? If so, how far beyond such a level need we go?[2]

If we pursue the piecemeal approach, how are we to bring a hypothetically entertained mathematical structure to bear on the material world? The most straightforward and general way is simply to continue employing second-order formulations as we have in the treatment of pure mathematics. This allows us to speak of classes of whatever non-mathematical objects we recognize and of relations between such objects and those of a hypothetically entertained pure mathematical structure. In such a framework, the representation of a great deal of applied mathematics is then quite straightforward.

To illustrate, let us consider a simple statement of numerical comparison, say, "There are more spiders than apes (and a definite finite number of each)." (The parenthetical clause is added so that some apparatus of natural numbers is required.) Using the second-order

---

[2] Similar questions can be asked with regard to constructivity of various sorts (e.g. predicativity). This has been a topic of intensive study by logicians (cf. e.g. Buchholz *et al.* [1981] and Feferman [1988] and references cited therein). Although we are not concerned with such questions here—having already invoked the power of full power set operations as integral to the modal-structural approach—we do see great interest in efforts to find the limits of varying degrees of constructivity, especially in connection with the mathematics employed in the physical sciences.

formalism of Chapters 1 and 2, with our language expanded to include the relevant non-mathematical predicates (in this case, just 'spider' ('$S(x)$') and 'ape' ('$A(x)$'), we can represent the statement by,

$\Box \forall X \forall f [\omega(X,f) \supset \exists \phi \exists \psi \exists n \exists m($"$\phi$ is a 1–1 correspondence between the class of all x such that $S(x)$ and the f-predecessors of n in X" & "ψ is a 1–1 correspondence between the class of all x such that $A(x)$ and the f-predecessors of m in X" & $m <_f n)]$,

(3.1)

where the clauses in quotes are written out in the obvious way and where $<_f$ is the strictly-less-than relation on X induced by f.

In ordinary language, we would read this modal conditional as a counterfactual: If X,f were any ω-sequence, there would be ... etc. However, there is a crucial proviso that must be understood: in entertaining an ω-sequence, it is assumed that such a structure does not interfere in any way with the actual material situation to be described. Since we shall have more to say about this, let us give it a name, say, *the non-interference proviso*. Without it, (3.1) would not express what is intended. If we imagine, for instance, a world in which an ω-sequence were made up out of distinct apes, the counterfactual would fail. What has gone wrong here is obviously that the wrong class of apes is being counted. One might seek to remedy this by employing an "actuality operator" before the conditions '$A(x)$' and '$S(x)$', to try to guarantee that it is the class of actual apes and that of actual spiders that are being compared.[3] The problem with this, however, is that we have still to guarantee that all the actual apes and spiders "show up" in any world hypothetically entertained. For, remember, we are not treating relations and functions as intensions, i.e. as operating "across worlds"; our second-order objects are treated as confined within any hypothetically entertained world. Thus, we cannot "move back" to the actual world and have a relation ϕ map actual objects to non-actual "numbers". Instead, we must *stipulate* from the outset that the only possibilities we entertain in employing the '$\Box$' are such as to leave the actual world entirely intact.

Of course, in most applications of mathematics, only a portion of the actual world is in question, and in such cases it would suffice to permit a broader interpretation of the '$\Box$', allowing worlds which

---

[3] For details on the formalism of actuality operators, see Hodes [1984].

differ from the actual even in material respects so long as such differences occur only outside the region of application. In such cases, there is no reason in principle why the atomic components of hypothetically entertained pure mathematical structures could not themselves be taken to be material objects of the same sort as occur actually. Moreover, in such cases, room can be allowed (literally) for further (mutually discrete) concrete objects to serve the role of ordered k-tuples of the hypothetical mathematical objects together with whatever actual objects are recognized. And then the second-order variables can be interpreted "nominalistically" in the manner of Chapter 1, § 6. Still, it will be necessary to stipulate that whatever "extra" material objects are entertained for such purposes are "causally isolated" from the region of actuality to which the mathematics in question is being applied.

Obviously there are limits to such an approach, since applied mathematics must also make room for cosmology, indeed for any science in which the large-scale structure of space-time is at issue. In such cases, it may be necessary to entertain "objects" as components of pure mathematical structures which are not themselves "in space-time". The options here are bound up with other issues concerning the "reality" of space-time, and, at this point, we wish only to alert the reader to the question. The topic will arise again below, at which point we shall have more to say on it.

Already it should be clear that even the most elementary applied mathematics on the modal approach is intimately bound up with conditions stipulating that at least part of (perhaps the whole of) the actual world be "held fixed", when reasoning about hypothetically entertained mathematical objects. So far, we have stated such conditions in rather general, global terms, bringing in explicit reference to the actual world or the actual condition or state of (some part of) the actual world. As terms such as "causally isolated" suggest, conditions of fixity or "non-interference", thus phrased, appear to embody some rather strong assumptions of "physical realism", especially the assumption that it even makes sense to refer to "the actual world", or "the condition of this system of physical objects", apart from any relativization to a language or theory or conceptual framework, etc. This raises one of the most interesting questions that an inquiry into applied modal mathematics uncovers: Is this apparent dependence of the cogency of applied modal mathematics on non-trivial assumptions of "physical realism" a genuine dependence, or can it in principle be eliminated? And, if elimination proves to be impossible, just what conclusion should be drawn as to the necessary commitments of the modal structural approach? We shall return to these questions below (§ 4), after having first dealt with the relatively more tractable issues concerning the strength of suitable mathematical frameworks.

Before proceeding, let us note a further assumption implicit in applied modal mathematics. Just as in the case of pure mathematics, there must also be axioms of modal existence—of the possibility of structures fulfilling the conditions of the antecedents of the counterfactuals. Without such axioms, of course, all counterfactuals with the antecedent in question could be vacuously true, and the translation pattern would break down. Here, and in what follows, it should be understood that the appropriate modal-existence postulate must accompany a fully explicit formalization of the applied modal theory. (It should by now be clear how to write out such postulates, and we will not stop to do so.) Due to the increasing uncertainties of such postulates as we move further from the realm of experience, there is a natural motivation—on the modal approach, as on the platonist—for seeking to carry out as much applied mathematics as possible within a minimal mathematical framework. Some aspects of this will be considered in the following sections.

Returning to illustrations, let us consider scalar magnitudes such as *mass* (say, non-relativistic, for simplicity). On standard platonist treatments, such a quantity is represented as a real-valued function defined on a domain of objects, either particles, or space-time points or regions, together with a given operationally specified unit. Suppose the worst, that each space-time location is to be assigned a real mass (or density). On the modal-structural approach, it suffices to entertain a single separable ordered continuum (as defined in Chapter 1, § 5), whose elements serve as "real numbers". These now can serve the double role of representing space-time points (via a pairing function which allows us to speak of ordered quadruples of reals in the usual way) and of representing the values of scalar quantities. A quantity such as *mass* is then a second-order object and can even be taken as a subset of "reals", each such coding an argument and corresponding value via the fixed pairing function. Thus, to represent a statement, ordinarily written as

$$m(x) = r,$$

where 'm' is a function constant introduced as abbreviating 'mass', we can write (following the notation of Chapter 1, § 5):

$$\Box \forall X \forall f \left[(\wedge \mathrm{RA}^2)^X \left(\begin{array}{c} \leq \\ f \end{array} \right) \supset \exists F (F \text{ "represents mass"} \right.$$

$$\left. \& \, F(x) = r \text{ in } X) \right], \tag{3.2}$$

where the clause in quotes must be spelled out as follows:

(i) *F* takes on (within limits of experimental accuracy) all actually measured values experimentally determined as values-of-mass;

(ii) *F* agrees (within experimental limits) with all theoretically predicted values-of-mass under real world conditions (the theory being Newtonian mechanics).

If our original singular statement is understood as a low-level empirical one—only loosely tied to a theory, as expressed in clause (ii)—, these conditions are probably adequate. However, if the statement is understood as part of an application of a whole theory—as it would be in any sophisticated application of mathematics—, then a further condition must be added to the effect that the mass-representing (mathematical) object satisfies the laws of the theory, or is part of a model of those laws. (Whether an intrinsic second-order statement of this over an RA^2 structure can be given, or whether ascent to richer structures is required, will depend on the detailed formulation of the theory in question.) Given such a condition, the utility of normal mathematical applications in permitting inferences as to further behaviour of the system in question will be accounted for, much as it is on familiar platonist (model-theoretic) treatments.

Intuitively, (3.2) can be read, "Were there any separable ordered continuum (non-interfering with the actual material world), there would be a mass-representing function assigning the value *r* (of the continuum) to point *x*", where 'mass-representing' is spelled out as suggested. Now it should be noted that the reference in these clauses to "measured values-of-mass" and "predicted values-of-mass" must be interpreted in terms of operational procedures and symbolic calculations. Since these conditions enter into the hypothetical conditionals designed to replace apparent reference to mathematical objects, clearly 'values' cannot be taken as referring to mathematical objects. (Hence the hyphens in 'values-of-mass'.) Rather, 'measured values' should be understood in terms of concrete "pointer readings", generally associated with certain symbols for real numbers. (More realistically, they would be associated with symbols for rational numbers; and, in some instances, some sequence of successively more and more accurate measurements may be specified for generating a real number.) Calculations may be involved as well (as they typically are in any sophisticated measurement procedure), and these generate number-symbols as well (e.g. decimal or binary representations, etc.). These then take on a definite meaning when a hypothetical "structure for the reals") is entertained in a conditional such as (3.2). (Together with any separable ordered continuum there is an associated correspondence between notations used in practice and the "points" of the continuum. This is induced by a representation of rationals within the continuum (cf. Chapter 1, § 5).)

Nor is this appeal to operational procedures and correspondences between notations and various ways of identifying real numbers and so forth a peculiarity of a modal-structural treatment. Standard platonist treatments of applied mathematics implicitly must invoke quite similar machinery, and must recognize a "relativity of reference" to "ways of taking" natural numbers, rational numbers, real numbers, etc. For the most common form of platonism, all such objects are set-theoretic constructions, and, of course, an infinite variety of these can serve the purposes of mathematical practice equally well. Within set theory, ordinary reference to a real number, say, is relative to a construal of the natural numbers as sets, to a pairing function, to constructions of negative integers and rationals, and to constructions of reals (e.g. via Dedekind cuts, or Cauchy sequences, etc.). The modal structuralist merely "does all this relativity one better" in dispensing with any actual mathematical objects at all in terms of which "reference to mathematical objects" is understood.

In any particular case, whether a hypothetically entertained mathematical object represents a physical magnitude is to some extent a vague matter, due to the need to take into account the approximate nature of measurement procedures in most scientific applications. This is, of course, reflected in the references to limits of experimental accuracy in the above clauses. This means that, in general, there will be multiple, extensionally divergent mathematical objects (functions) that qualify equally well as representations of a magnitude. Whether, for instance, to represent the path of an object through space as a continuous function of time or as, say, a piecewise continuous one, or even a highly discontinuous one, or whether even to

represent time as a continuum in the first place, are matters under-
determined by any direct experimental procedures. Thus, choices
must be made on other grounds, and probably the most decisive
grounds in many cases are considerations of simplicity and
convenience—considerations such as that a well-developed mathematical theory of continuous functions exists enabling us to perform
vital calculations, and that, on such practical grounds, we seek a
theory couched in mathematical terms that we can handle. (As
Chihara [1973] brings out, it is just this slack that raises the
prospects of constructivist substitutes for classical applied mathematics.) This, in turn, raises thorny questions concerning the conven-
tionality of our scientific theories, questions that certainly cannot be
resolved here. In my own view, it seems obvious that any sophisticated application of mathematics to the material world involves a significant degree of idealization, which implies a significant degree of
conventional choice based on pragmatic considerations. At the same
time, this by no means undermines the "objectivity" of our scientific
theories, provided that that "objectivity" is properly understood.
While, in general, we cannot say that there are such and such magnitudes in nature represented precisely within a unique mathematized
theory, we still may be able to say that *nature is such as to permit* representation within a *range* of mathematical models, and that this
range *includes* such and such mathematically precise description.
Perhaps this is all the "objectivity" we ever require. In any case, without pursuing this further here, suffice it to say that *any* thorough
account of applied mathematics must at some stage come to grips
with these questions. They are by no means peculiar to a structuralist treatment.

With these essentials of the approach in mind, let us now turn to
the question of how rich, mathematically, our hypothetical structures need to be in order to support applications of our best modern
physical theories.

§2. Carrying the Mathematics of Modern Physics: RA² as a Framework

As has already been indicated, the RA² framework is known to be a
very powerful theory with regard to the requirements of applied

mathematics.[4] Moreover, as will be brought out below, there is a sense in which it defines a "limit of nominalism", a limit to the mathematical richness of what can be conceived of as "concrete structures". Thus, there are special reasons for focusing on the representing powers of RA^2. Can it really do justice to modern physical theories, especially General Relativity and Quantum Mechanics? A full-scale treatment of this would take us too far afield; but a brief glimpse will suffice to call attention to some of the fascinating issues that arise in this area.

In the case of General Relativity, matters are complicated by the fact that the theory is standardly presented in two very different ways. On the one hand, there is the "extrinsic" presentation—familiar to physicists—in which everything is carried out explicitly in terms of coordinate systems and transformations among them; on the other hand there is the "intrinsic" or "invariant" presentation in terms of abstract geometric objects making no reference to coordinate systems.[5] Now, it is the intrinsic presentation that is mathematically more elegant and, moreover, can be argued to provide a clearer idea of the content of the theory and of such matters as how it compares with other space-time theories (e.g. Newtonian gravitation). However, ordinary physical applications make use of coordinate systems, so that a representation of the extrinsic presentation may be regarded as adequate for most purposes.

So long as we remain with the extrinsic presentation, there is little doubt that the system RA^2 is powerful enough to express and derive what is normally required. (That is, say, standard texts could be systematically translated into RA^2 and all results derived.) Geometric objects—vectors, tensors, etc.—are treated in terms of their components in a coordinate system and the rules for transforming them to other admissible coordinate systems. Coordinate systems can be viewed as 1–1 maps from regions of space-time to $\mathbb{R}^4$; and the components of a geometric object are given by suitably continuous real-valued functions. (For example, the components of a tangent

---

[4] As stated in Burgess [1984], p. 386, it is "probably sufficient to develop, making much use of coding devices, all the mathematics that has found scientific applications up to the present".

[5] The intrinsic, coordinate-free approach goes back to work of Cartan [1924]; it was employed in the study of Newtonian gravitation by Trautman [1966]; and it is presented in modern sources such as Misner *et al.* [1973]. For a readable presentation of the once more common coordinate based formulation, see e.g. Ohanian [1976]. For a comparison of the two from a foundational perspective, see Friedman [1983], Ch. 2.

vector field T_σ to curve σ ($\sigma = \sigma(u)$ being a smooth map from an interval of $\mathbb{R}$ to space-time), relative to coordinates $\langle x_i \rangle$ consist in four functions, $\dfrac{\mathrm{d}(x_i \circ \sigma)}{\mathrm{d}u}$, $i = 0, 1, 2, 3$.) Any such function can generally be coded by a single real, being determined by its behaviour on a countable subdomain.[6] (Think of the simplest case of a continuous function from $\mathbb{R}$ to $\mathbb{R}$: a real can code its behaviour at rational arguments. Even countably many discontinuities can be allowed. In the case of, say, tensor fields on (a region of) space-time, a countable subdomain of $\mathbb{R}^4$, as described in a given coordinate system, suffices to determine the field. So long as we remain within a single coordinate system, it makes no difference whether we regard the field as defined on (part of) space-time or on (part of) $\mathbb{R}^4$ itself, via the coordinate functions. In an invariant, coordinate-free presentation, however, the distinction becomes significant, and in some cases leads to greater abstractness, since we then must in effect keep track of all coordinate systems at once.) The transformations determining, say, a general tensor operate on finitely many component functions, hence, by coding, on finitely many reals, and yield reals coding transformed functions as values, clearly within RA^2. And, as just suggested, a (suitably continuous) tensor field can be described, relative to a coordinate system, by a function which gives the component functions as values on a countable dense subdomain of $\mathbb{R}^4$. Hence, a tensor field can be coded as a single real number! And all the usual operations on such fields, including covariant differentiation, can be introduced as functions from reals to reals, within RA^2 (which, recall, includes the full second-order comprehension scheme). This much should at least make plausible the claim that all the mathematics actually required in any ordinary physical application of General Relativity can be carried out without transcending third-order number theory (equivalently RA^2). (And, by making sufficient reliance on "approximating functions", a great deal can probably be carried out in a predicative subsystem of analysis, i.e. of PA^2.)[7]

For ordinary applications, the story could end here. However, not all applications need be "ordinary". This forces us to raise a difficult

---

[6] Here and below, any reference to "the reals" ("the natural numbers", etc.) is to be understood as convenient abbreviation for the obvious more long-winded modal-structural substitute, framed in terms of hypothetical RA^2 (PA^2, etc.) models.

[7] For an interesting development of this idea as a key step toward a "modal nominalism", see Chihara [1973]. See also Burgess [1983].

question: is the intrinsic formulation really dispensable in favour of the extrinsic for all scientific purposes? Even if all the usual sorts of applications involving specific calculations can be carried out in RA^2 or in some weaker system, this does not settle the matter, for there are questions of theoretical importance that go beyond such applications, but which a mathematical framework ought to be capable of representing if it is to do justice to the "scientific enterprise". As a case in point, relevant in the present context, consider the whole issue of "relativity principles" and requirements of "general covariance", thought by many (including Einstein) to distinguish General Relativity from flat space-time theories. Remaining at the level of coordinate-based formulations, one can readily be misled into thinking that General Relativity differs from flat space-time theories in satisfying such a principle, since one considers transformations among "general curvilinear coordinates" rather than a privileged class of "inertial systems". However, if one considers intrinsic formulations, it becomes evident that this is mistaken, and that the demand of "general covariance"—that dynamical laws retain "their form" under arbitrary transformations among coordinate systems—really comes to nothing more than the demand that those laws be given an intrinsic coordinate-independent formulation, something that is possible for flat as well as curved space-time theories.[8] In fact, if one looks at space-time theories model theoretically, one sees that extrinsic formulations—in terms of equations involving coordinates—pick out a well-defined class of models only relative to a choice of coordinates. In a different system of coordinates, the same differential equations (e.g. one looking like a geodesic equation) will pick out a different class of geometric objects (e.g. tangent vector fields), hence a different class of models.[9] Intrinsic formulations automatically overcome such problems.

Now, if we understand "applied mathematics" broadly enough to

---

[8] Cf. Friedman [1983], Ch. 2, §2, and sources cited therein. See also Hiscus [unpublished].

[9] This point is made clearly in Friedman [1983], pp. 50–1. In a coordinate-free presentation, the geodesic equation takes on the form

$$D_{T_\sigma} T_\sigma = 0,$$

in which D is an affine connection (covariant derivative operator) and T_σ is a tangent vector field (to curve σ). If the manifold is flat, in an inertial coordinate system $\langle x_i \rangle$, this equation takes the form,

$$da^i/du = 0,$$

include the theoretical insight that such comparisons yield, whatever mathematics is required in the abstract intrinsic formulation cannot readily be dismissed as dispensable. Granted, this is an unusually broad interpretation of "applied mathematics". But the theoretical understanding in question is at the heart of the sciences, and, if a "dispensability argument" (to the effect that mathematical structures richer than X are not "needed" to "do natural science") is to carry philosophical force, such considerations must be taken into account.

Thus, we cannot avoid considering the problem of representing the mathematics of the intrinsic presentation. While the matter cannot be definitively settled here, the situation seems to be this: the abstract theory of manifolds transcends the RA^2 framework, but essentially only at the earliest stages, namely in the abstract characterization of manifolds themselves. Once given a manifold, it appears that, in fact, with sufficient reliance on coding devices, the system RA^2 is capable of representing the rest of the mathematical superstructure of abstract differential geometry employed in the intrinsic presentation of General Relativity. Moreover, a great many particular manifolds actually encountered in space-time physics can be introduced explicitly in RA^2, making use of second-order logic.

The intrinsic formulation begins with the idea that, at least locally, space-time has the structure of an 4-dimensional smooth (C^∞) manifold. An n-dimensional C^∞ manifold consists in an arbitrary nonempty set M together with a maximal system of charts—1–1 maps from subsets U of M to open sets of $\mathbb{R}^n$—suitably interrelated so as to induce the "local smoothness structure" of Euclidean n-space on M.[10] More precisely, an *n-chart* is a pair (U, f) where $U \subseteq M$ and f is a 1–1 map from U onto an open set of $\mathbb{R}^n$; an *n-subatlas* on M is a family of n-charts such that (1) they cover M, i.e. the union of the domains of the charts is M; (2) for any two distinct points p, q of M

but in a non-inertial coordinate system, $\langle y_i \rangle$, the same invariant equation takes on the form,

$$da'^j/du + \text{"correction terms"} = 0,$$

where the "correction terms" have the form

$$\frac{\partial y_j}{\partial x_i} \frac{\partial^2 x_i}{\partial y_k \partial y_l} a'^k a'^l,$$

in which the a' are components of the tangent vector expressed in the non-inertial system.

[10] Here and below our discussion is based primarily on Hicks [1971] and occasionally also on Malament [unpublished].

there are charts (U_1, f), (U_2, g) with p in U_1, q in U_2 and $U_1 \cap U_2 = \varnothing$ (Hausdorff property); and (3) any two charts (U_1, f), (U_2, g) of the family are *compatible*, meaning that $f \circ g^{-1}$ and $g \circ f^{-1}$ are smooth (C^∞), whenever defined on open domains. Finally, an *n-atlas* is obtained from an *n*-subatlas by adding all *n*-charts compatible with all those of the *n*-subatlas. M together with an *n*-atlas on M is an *n*-dimensional C^∞ manifold.

Note that, while we could perfectly well take M to be $\mathbb{R}$, i.e. to consist of points "labelled as real numbers" for the purposes of a coding in a logical formalism (while still abstracting from any undefined metric or topological structure), and while we can always code the composite maps $f \circ g^{-1}$, etc. as reals (being determined by their behaviour on a countable dense subset of $\mathbb{R}^n$), we have no means of coding the f of the charts as reals. Charts are second-order RA^2 objects. This makes maximal families of charts (atlases) essentially third-order over the reals, which is why the general notion of manifold transcends RA^2.

Moreover, when we develop calculus on manifolds, it appears that we are beyond RA^2 once and for all, for, once we leave coordinates behind, the very notion of a vector becomes apparently too abstract: A vector is standardly taken as a derivative operator ("derivation"), i.e. as a map γ from all real-valued smooth functions about a manifold point m to the reals, meeting the requirements of linearity ($\gamma(f + g) = \gamma(f) + \gamma(g)$, $\gamma(af) = a\gamma(f)$) and the Leibnizian property ($\gamma(fg) = \gamma(f)g(m) + f(m)\gamma(g)$). As such vectors are third-order objects over the reals, beyond RA^2, and then tensors and more general derivative operators are at least as abstract. However, without appealing to coordinates, we can equivalently take vectors to be tangent vectors to smooth curves σ (from a connected interval I of $\mathbb{R}$ to the manifold domain M). Given a smooth curve σ and $s_0 \in I$ with $\sigma(s_0) = p$, a tangent vector $\sigma|_p$ to σ at p can be introduced via

$$\sigma|_p(f) = \frac{\mathrm{d}}{\mathrm{d}s}(f \circ \sigma)|_{s_0},$$

for all smooth real-valued functions f on a neighbourhood of p. (N.B. The notions of smoothness for maps from $\mathbb{R}$ to M or M out to $\mathbb{R}$ are introduced via the smoothness of composite maps from $\mathbb{R}$ to $\mathbb{R}^n$ or from $\mathbb{R}^n$ to $\mathbb{R}$ given by the chart functions and their inverses.) Now, $f \circ \sigma$ is a smooth function from $\mathbb{R}$ to $\mathbb{R}$, and thus can be coded by a real. Moreover, the derivative operator on the right is continuous so it, too, can be coded by a real. Thus, abstract vectors are brought down to the level of first-order RA^2 objects. (The proof that every original abstract vector can be taken as a tangent vector to a smooth curve in M is standard, but, unfortunately for the RA^2 reductionist, it cannot be stated in RA^2.)

But now, tangent spaces, dual spaces, tensors, tensor fields, covariant derivative operators—in sum, all the further apparatus needed to carry out the intrinsic formulation of General Relativity—come within the scope of RA^2. The tangent space V_m at a point m of the manifold is the set of all tangent vectors at m, and is a vector space (of dimension n) over the reals. It can be taken as a set of reals via the coding of vectors just indicated. The dual space $V_m{}^*$ of linear real-valued maps on V_m is then also identifiable as a set of reals: each ϕ in $V_m{}^*$ is determined, by linearity, by its action on finitely many basis vectors in V_m (which can be chosen arbitrarily from any coordinate system about m); hence ϕ can be coded by a real. Next, a general tensor—as a multilinear map T from finite cross-products of the form $V_m \times V_m \times \ldots \times V_m \times V_m{}^* \times V_m{}^* \times \ldots \times V_m{}^*$ into $\mathbb{R}$—is determined by its action on basis vectors in the component spaces making up the domain of T; hence T itself can be coded as a single real. Thus, smooth vector and tensor fields—maps assigning vectors or tensors, respectively, to points of open subsets of M—come within the purview of RA^2, as sets of reals. (Here we make use of real labels of the points of M. Then a field can be coded as a set of ordered pairs of reals, hence as a set of reals.)

With one more step we have essentially all that is needed: we must have a way of talking about (quantifying over) covariant derivative operators, or "connections" on M. Such a connection D is introduced as an operator assigning to C^∞ fields X and Y, with a domain A, a C^∞ field $D_X Y$ with domain A, obeying four conditions (ensuring appropriate linearity and Leibnizian behaviour). Prima facie, such operators are third order over the reals, and beyond RA^2 by one level. However, the conditions on D imply that it is uniquely determined in any open domain by its action $D_{e_i} e_j$ on a fixed finite base field $e_1 \ldots e_n$ of independent C^∞ vectors.[11] Since the e_i (as fields) are

---

[11] See Hicks [1971], p. 57. The four conditions on D are

    (1) $D_X(Y + Z) = D_X Y + D_X Z$
    (2) $D_{(X+Y)}Z = D_X Z + D_Y Z$
    (3) $D_{(fX)} Y = f D_X Y$
    (4) $D_X(fY) = (Xf)Y + f D_X Y.$

coded as sets of reals, this means that, on the domain of these fields, D can be represented as a set of reals coding a three-place relation (that is, $e_i \times e_j \times D_{e_i}e_j$). Now, if one assumes (as one usually does in the context of General Relativity) that the manifold M is *separable*— i.e. that it can be covered by countably many chart domains—then countably many such three-place relations, codable as a single such, suffice to represent D throughout the manifold M. In this manner, even quantification over connections is brought within the scope of RA^2.

Still, remarkable as all this may be, we are not able to state—much less prove—in RA^2 fundamental general theorems on manifolds, such as the theorem that there exists a unique Riemannian connection on a (semi-) Riemannian manifold, or the theorem that a metric tensor g_{ab} on a manifold determines a unique general derivative operator compatible with this metric. For recall that the general notion of manifold is not available in RA^2. The best we can do is introduce particular manifolds—e.g. the manifold $\mathbb{R}^n$, or the *n*-sphere manifold, etc.—by explicitly axiomatizing a system of charts. (For instance, in second-order logic, we can write down an explicit definition of the predicate "is a chart of the $\mathbb{R}^n$ manifold atlas": we simply specify the identity functions on open sets of $\mathbb{R}^n$ (as a point set with the usual topology), which gives a C^∞ subatlas; then we specify that any chart compatible with all those of the subatlas are "charts of the $\mathbb{R}^n$ manifold atlas".) Again, for ordinary applications, such procedures are probably adequate. But without fundamental theorems on differentiable manifolds of the sort mentioned, we can hardly claim to do justice to the intrinsic viewpoint.

From these, it follows that if X_m (X at point m of the manifold) $= \sum_1^n a_i(m)(e_i)_m$ and $Y = \sum_1^n b_j e_j$ on the domain of the base field $e_1, \ldots, e_n$ (intersected with the domain of Y), then

$$(D_X Y)_m = [D_X(\sum b_j e_j)]_m = \sum_j [(X_m b_j)(e_j)_m + b_j(m) \sum_i a_i(m)(D_{e_i} e_j)_m].$$

Thus, $D_X Y$ is fully determined by the $D_{e_i} e_j$ together with the coefficients of X in the given base, the coefficients of Y, and the $X_m b_j$ (the directional derivatives of the Y components along X_m). In sum, the action of D on the base fields allows computation of $D_X Y$ in terms of the behaviour of Y on a curve that fits X.

Incidentally, it should be noted that, in appealing to a basis for the purpose of coding derivative operators or other objects, we are not thereby abandoning the invariant formulation. The coded objects are still invariant objects and they appear as such in the theoretical equations governing them, that is, those equations are stated independent of any coordinate system.

Our discussion, thus far, has presupposed, for the sake of argument, that various coding devices are legitimate in reducing mathematical machinery to low levels of abstraction over the natural numbers. However, it should be recognized that delicate problems of justification arise in connection with such devices. These pertain to the status of the mathematical knowledge employed in the introduction of a coding in the first place, e.g. when one says that an abstract operator is uniquely determined by its action on finitely many or countably many arguments of a certain sort, and therefore can be represented by, say, a real number or a set of reals, etc. In what sense of "could" *could* a mathematical physicist carry out all relevant derivations and calculations within the coding framework, without ever stepping outside in order to be reminded of "what on earth is really going on"? Since our non-reductionist conclusions can be based on the more decisive case that, even *with* coding, relevant mathematics goes beyond even the full power of RA^2, we have preferred to develop that case without examining the delicate epistemic problems raised by appeals to coding. But, we believe, those problems are genuine and deserve further investigation.

When we come to Quantum Mechanics, the situation is at least as problematic as in the case of General Relativity. In many ordinary applications, a greal deal of the mathematics can certainly be carried out within the framework of RA^2, but once we consider more theoretical and foundational matters, we seem to require more abstract structures. In standard cases, quantum states can be represented as square-integrable complex-valued functions on an underlying real space, and moreover a countable collection of continuous functions serves as a basis in the Hilbert space of such functions.[12] Thus, arbitrary quantum states in such a space can be represented by a countable sequence of basis functions, each codable as a real, hence by a real. Linear operators on such functions are then, prima facie, at the level of second order RA^2 variables, as are the (closed) subspaces of the Hilbert space (identifiable with the projection operators). If we now consider probability measures—countably disjointly additive [0–1]-valued functionals on the subspaces of the Hilbert space representing the system (where, here "disjointness" of subspaces means they are orthogonal)—we have, prima facie, climbed past

---

[12] For examples and proofs of key basis theorems, see e.g. Prugovecki [1981], Ch. 2.

RA². However, given separability of the Hilbert space,[13] each sub-space S can itself be identified with a countable collection $\{f_i\} = C$ of basis vectors such that C spans S and C is dense in S. And, given that each of the f_i is codable as a real, so is each subspace. And then, any probability measure can be represented as a function from reals to reals, i.e. at the second order in RA².

In fact, due to an important theorem of Gleason [1957], all measures on the subspaces of a Hilbert space of dimension ≥ 3 are given by the quantum mechanical algorithm, that is, they are induced by the pure and mixed quantum states together with the usual rules for calculating probabilities. Hence, again under the above assumptions (separability of the Hilbert space and real codability of basis functions), the measures themselves can be taken as reals (i.e. coding the density matrices which represent the pure and mixed states on the usual presentation of the theory). Thus, reasoning involving probability measures in the vast preponderance of ordinary applications of quantum mechanics can be carried out in RA², and probably even in theories considerably weaker than RA².

However, in order to arrive at this conclusion, we assumed that the Hilbert space representing the physical system be separable. And, furthermore, to take advantage of Gleason's theorem (permitting the representation of measures as reals rather than sets of reals), we implicitly made use of enough mathematics to prove Gleason's theorem. In fact, the proof of Gleason's theorem serves as a nice illustration of the potential physical significance of mathematics that transcends RA².

As a theorem about probability measures on the subspaces of *separable* Hilbert spaces (of dimension ≥ 3), Gleason's theorem can surely be proved in RA². All the hard work in the proof takes place in Euclidean three-dimensional space; moreover, every infinite dimensional separable Hilbert space is isomorphic to ℓ^2, the space of infinite sequences $(x_1, x_2 ..., x_k, ...)$ of complex numbers such that

---

[13] A Hilbert space, H, is separable iff it possesses a countable collection of vectors, $f_1, f_2, ..., f_n, ...$, which is dense in H, i.e. for any f in H and any $\varepsilon > 0$, there is an f_i such that

$$\|f - f_i\| < \varepsilon,$$

where the norm $\| \ \|$ is defined by the scalar product on H via

$$\|g\|^2 = (g, g).$$

For further details on the mathematical formalism of quantum mechanics, see e.g. Jauch [1968].

$\sum_{k=1}^{\infty} |x_k|^2$ is finite, with the inner product of (a_k) and (b_k) given by $\sum_{k=1}^{\infty} a_k^* b_k$. (This fact can be proved in RA^2 itself, making use of an intrinsic second-order statement of what it is to be an infinite dimensional separable Hilbert space, on exactly the plan of the second-order logical description of mathematical structures we have been using throughout. Second-order logic suffices here, taking vectors in a Hilbert space to be first-order objects.) And everything we need to say about measures on the subspaces of ℓ^2 can be said within RA^2 along the lines already indicated.

However, what happens if the assumption of separability is dropped? The above representation via coding then clearly breaks down. Yet, there is a generalization of Gleason's theorem, proved in an abstract setting, which applies to non-separable as well as separable Hilbert spaces.[14] This theorem proves the existence of a unique function $w(p, b)$ defined on the atoms p and propositions b of an arbitrary quantum proposition system (represented by the lattice $\mathscr{S}(H)$ of closed subspaces of an arbitrary Hilbert space of dimension ≥ 3), where $w(p, b)$ satisfies:

   (i) $0 \leq w(p, b) \leq 1$,
   (ii) $p < b$ iff $w(p, b) = 1$,
   (iii) $b \perp c \Rightarrow w(p, b) + w(p, c) = w(p, b \vee c)$.

Since p is an atom, it can be represented by a (normalized) vector f in H. It then can be proved that the unique function satisfying these conditions is given by

$$w(p, b) = \phi(P^b(f), f), \tag{*}$$

where P^b is the projection operator corresponding to the proposition b, and ϕ is the definite bilinear form (inner product) on the Hilbert space H. (Later, when one specializes to separable Hilbert spaces, the usual representation of $w(p, b)$ in terms of the trace and a density matrix is recovered.)

Now, the point is that this generalized version of Gleason's

---

[14] See Piron [1976], pp. 73–81. A prelude to this is an elaborate representation theorem to the effect that an arbitrary quantum proposition system can be represented by a family of lattices of closed linear varieties (subspaces) over abstract Hilbert spaces, which need not be separable. What would serve as a minimal framework for proving this representation? Since its physical significance is unclear, we have preferred to concentrate on Gleason's theorem, some of whose physical content is easier to specify. Incidentally, this source provides a quite readable proof of Gleason's theorem.

theorem has an indirect physical significance. As an important corollary to Gleason's theorem, there can be no dispersion-free measures on the subspaces of Hilbert space, as sought by (non-contextual) hidden-variables programmes.[15] Now this is already implicit in the generalized representation (*) of $w(p, b)$ above (since, for any p, b can be varied so that $w(p, b)$ takes on values strictly between 0 and 1). If a restriction to separable Hilbert spaces is built into the very statement of Gleason's theorem—a statement that one succeeds in proving in a restricted framework such as RA^2 or some weaker theory—it might appear that hidden variables could be reinstated by complicating the Hilbert space (i.e. enlarging its dimension, in this case). (The standard examples of Schrödinger wave mechanics could then perhaps be viewed as simplifying approximations.) Thus, once again, if mathematics of importance in understanding the physical theory is included within "applied mathematics", we find ourselves moving to rather abstract settings, beyond the reaches of RA^2.

But why, it may be asked, should one dwell on RA^2? Is it of anything more than purely technical interest that modern physics at certain junctures takes us past this level? Yes, for there are implications concerning the "nature of the objects" that mathematical physics must entertain, even if only hypothetically. So long as we remain at or below the level of RA^2, models of our mathematical theories can be conceived as concocted out of objects that qualify as "concrete" in the sense of "occurring in, or part of, physical space or space-time", assuming, of course, that we conceive of physical space and space-time as continuous, as we generally do. Any connected open interval of such a space can serve as the domain of a model of RA^2, i.e. its points serve as the ground-level objects of the mathematical theory. And the second-order monadic quantifiers can be understood as

---

[15] A non-contextual hidden-variables theory can be understood as demanding dispersion-free measures on the subspaces of Hilbert space, i.e. measures representing states which assign probability either 0 or 1 to every statement of the form "quantum magnitude A (pertaining to the system represented by the given Hilbert space) has a value in Borel set S". Such a hidden-variables programme is called non-contextual because it treats the quantum magnitudes (represented by linear Hermitian operators on the Hilbert space) as they stand, without relativizing them to experimental context (or to maximal compatible sets of operators, etc.). It is an immediate corollary to Gleason's theorem that, if the Hilbert space has dimension ≥ 3, there can be no dispersion-free measures (as quantum states always exhibit dispersion for some magnitudes, as they respect the Heisenberg uncertainty relations). For further information on the topic of hidden variables and no-hidden-variables proofs, see e.g. Belinfante [1973], Clauser and Shimony [1978], and references cited therein.

ranging over subregions of the domain. As noted above (Chapter 1, §6), this remains within the spirit of "nominalism" in the quasi-formal sense of Goodman [1977], Field [1980], and others. Moreover, invoking coding devices, one can also accommodate second-order polyadic quantifiers (as ranging over sums of atoms—points—coding k-tuples of first-order objects). Even if we do not countenance space-time points or regions as actual—even if we eschew "substantivalism" in one sense—we may with some plausibility claim to understand what it means to entertain such objects hypothetically, and this is all that is required for mathematics. Thus, if all the mathematics needed for physics could be represented at the level of RA^2, a kind of nominalism could probably be sustained. If space-time substantivalism (in the above sense) is upheld, the nominalism could be framed without the modality of mathematical possibility; if not, at least there would be a case for this version of "modal nominalism". (Of course, one need not insist that such "nominalist substitutes" for ordinary mathematics be employed in actual practice; the claim, rather, would be that, in principle, the limited means are all that are really needed, so that practice transcending those means need only be justified on pragmatic grounds, i.e. the transcendent mathematics *could* be treated instrumentally.)

However, if, in fact, as it presently appears, RA^2 is *not* adequate after all, then even modal nominalism in this sense is doomed.[16] It seems likely that structures for even richer theories need to be entertained, and with these we will have transcended what can even be conceived as part of space-time as we understand it. Note, moreover,

---

[16] There is still the possibility that physical theories can be "nominalized" within a purely relational ("synthetic") space-time framework employing the methods of Field [1980]. We shall have more to say about this below (§4). But note here that Field worked out a single example—Newtonian gravitation—and, as we noted above, in this case, space-time points and regions suffice for representing the mathematics of the theory. In fact, as our remarks above imply, it is not even necessary to introduce a purely relational space-time version of the theory; one can reinterpret the mathematics directly via coding. And then one bypasses the sort of representation theorem that Field highlighted. (The use of such a theorem confronts technical difficulties, to be reviewed below, §4.) Moreover, one respects the mathematics and need not adopt an instrumentalist stance. But—more important in the present context—there seem to be serious obstacles in the way of Field's programme as a strategy for nominalizing modern physical theories, especially quantum mechanics, in which the domains of the models are already highly abstract, and which do not lend themselves readily to a space-time reformulation. (In this connection, see Malament [1982].) Whether the programme can be made to work even for non-flat space-time theories, e.g. General Relativity, is, I believe, an open question.

that imagining higher-dimensional physical spaces,[17] coherent as this may be, does not touch the problem unless the cardinality of the dimensions is itself uncountably infinite (!), assuming the spaces (as domains of points, apart from further structure) are obtainable from the continuum by taking Cartesian products. (This is what one usually understands by "higher dimensional space".) Where c is the cardinality of the continuum, $c \times c = c$, as does $c \times c \times c$, etc., as does $c \times c \times \ldots \times c \times \ldots$, where the iteration is countably infinite. (In ZFC one proves this: $c^{\aleph_0} = (2^{\aleph_0})^{\aleph_0} = 2^{\text{card}(\aleph_0 \times \aleph_0)} = 2^{\aleph_0} = c$.) But the cardinality of the next full-level past RA^2, i.e. of the power set of $2^c = 2^{2^c}$, is not approached by the totality of second-order objects of such product spaces, which is, of course, still of cardinality 2^c. At any rate, the spaces it is necessary to entertain just to make enough distinctions—once we ascend beyond RA^2—bear little relation to the spaces of experience. We can hardly call such a framework "nominalistic" (unless, perhaps, we allow infinite stretching of words without change of meaning!).

Does this mean that the modal approach collapses into set theory after all! No, that by no means follows. Structures beyond RA^2 may not qualify as nominalistic, but we can still entertain such structures hypothetically, dropping any claim to "grasp them" by means of "geometric intuitions". Moreover, we may still adhere to the structuralist insight that the identity of the objects of such structures is of no interest to mathematics. We may still reason about such structures without benefit of any assumption as to "the nature" of the objects. Instead of "modal nominalism", then, perhaps we should speak of "modal *neutralism*". We retain, of our ordinary notion of object, only what is necessary for mathematical reasoning (pure and applied). In pure mathematical reasoning, it is necessary to assume only logical distinctness of the objects (i.e., distinctness in the sense of $\neq$) and the structural relations spelled out in axioms (which includes collectability and relatability expressed in second order comprehension axioms). In applied mathematical reasoning, it is, as we have already seen, also necessary to invoke an assumption of causal inertness or independence of the material world (usually just part, but sometime all, of it). Just what is involved here will be taken up below, after we have considered the matter of global frameworks.

---

[17] As was suggested in Putnam [1967].

§ 3. Global Solutions

On this approach, one seeks a framework for applied mathematics
that is so rich that we can be confident in advance that no physics—
past, present, or future—would require a yet richer framework. In
some sense, one would also like such a framework to be minimal. But
here one must be careful. Suppose it turns out that an optimal
physics (supposing just for argument's sake that there is such a
thing) for the world we occupy requires exactly 37 applications of the
power set operation past the first infinite level (i.e. the mathematics
of the theory could be formalized in 38th-order number theory). A
reasonable response would be, "So what?" Does this in any way cast
doubt on mathematics that can be formalized only at the level of
39th-order number theory? The cut-off seems arbitrary on two
counts: first, no new fundamental mathematical idea is involved in
simply iterating the power set operation one more time. Unlike the
leap from RA^2 to higher levels, no genuine matter of principle is involved. And second, even if 39th-order number theory is dispensable
in carrying out the optimal physics of this world, this is of little philosophical interest, since it is too easy to imagine that the physical
world might have been slightly more complex, in which case 39th-
order number theory would have been "indispensable". If "dispensability" of higher-order mathematics is to have any force at all, it
must not be of the character of a "lucky accident"! The force in question, I take it, is to persuade us that a formalist or instrumentalist
attitude toward the higher levels is tenable, that one need not conceive of the sentence pertaining to such levels as true or false. What I
am suggesting is that the "need" here must not be interpreted too
narrowly, on pain of trivializing the claim of dispensability.

It may be that these two considerations really boil down to one,
namely to the first one. If no new mathematical principle is at
stake—if, say, one is entertaining structures obtained simply by
repeating a given operation one more time, and this does not involve
fundamentally different sorts of objects (as the leap beyond RA^2
arguably does)—then suddenly adopting an instrumentalist stance
will seem arbitrary. Since we are invoking nothing essentially new,
we may claim more readily to conceive that a more complex physics
reigned.

In any case, it would be futile to seek a precise characterization of
"essentially new" mathematical principles. The best we can hope to

do is find particular examples which might then serve in defining a minimally adequate framework for applied mathematics.

A natural suggestion here would be minimal full models of Zermelo set theory, with urelements. Apart from the urelements, these are of the form $R_{\omega \times 2}$, i.e. the structure that results from arbitrary finite iterations of the power set operation beyond level ω. (If we have infinitely many urelements, we could simply allow finite iterations of power set above the urelements themselves. But we wish to make no such assumption. We may take the first infinite level to be independently motivated, say, along the lines of Chapter 1, §2.) But, in such models, only one limit level is recognized. A "new mathematical principle" that would generate further limit levels (without end) is, of course, Replacement. But, for applied mathematical purposes, the claim would be that this is unnecessary.

There is no difficulty in characterizing the relevant sorts of structures in the second-order logical framework we have been employing. One writes the Zermelo axioms in second-order form, i.e. *Aussonderung* is written as,

$$\forall P \forall a \exists x \forall y (y \in x \equiv P(y) \ \& \ y \in a). \tag{Auss}$$

To the Zermelo axioms, one then adds as a further axiom the statement,

$$\text{"There is just one limit ordinal",} \tag{R}$$

so that the height of any model is $\omega \times 2$ (as it would be standardly described in set theory). ('R' is for 'restriction'.) Call the resulting finite list of axioms Z^+. Then applied mathematical statements can generally be given the form,

$$\Box \forall X \forall f \left[(\wedge Z^+ \ \& \ U)^x \binom{\in}{f} \supset A \right], \tag{3.3}$$

where 'U' stands for the statement that certain non-mathematical objects of interest are included in the urelement basis of the structure, and where 'A' stands for a statement of application describing what would obtain concerning relations between the urelements and items of X relevant to the particular theory being applied.

There are a number of advantages to this global approach. First, (3.3) gives a uniform pattern that works independently of the particular applied theory. Second, the theory Z^+ seems tailor-made for applications typically encountered in mathematical physics, in which one begins at some infinite level (such as the reals or the complexes) and then freely moves up to higher-type objects, but always by a finite number of steps. Thus, the mathematical constructions actually encountered can be taken over intact; one need not resort to special coding procedures or other substitutes, and one then bypasses thorny questions of justifying such devices. (As was asked at the end of §2, when a coding is used to reduce the "level of indispensability", in what sense of "could" is one claiming that the coded mathematics *could* be used for all the purposes of the uncoded?) Third, one now can express statements about arbitrary "sets" of urelements. One has available the collecting operation of any model of Z^+ and is not confined to instances of second-order logical comprehension. Thus, we can represent directly the statement, for example, that every collection of urelements is at most countably infinite (if we wish), or that the maximum distance between urelements of any collection is finite, etc. The primary disadvantage would seem to be that, with (3.3), we give up hope of even modal "nominalism" in the sense discussed in §2. Instead, we have a modal neutralism, as already suggested. But, very likely, this is the best we can achieve within a realist, non-constructive framework anyway.

An important question, however, may be raised regarding (3.3) from the opposite end: granted that as much mathematics as can be formulated within Z^+ is to be included in a framework for applied mathematics, is this really enough? Might we not require, for example, the full power of ZF, or ZFC, or even ZFC + large cardinal axioms? Just because we haven't encountered the need for such stronger axioms yet doesn't mean that we won't, much less that we couldn't (in some suitable sense of "natural possibility"). The most that can be said for (3.3) is that it is adequate for known and foreseeable applications, but that defines no principled demarcation. In particular it should by no means be inferred that mathematical principles beyond Z^+ are generally dispensable from a scientific standpoint and should therefore be viewed in purely formalist terms.

Now, it may seem on the face of it outlandish that large cardinal axioms, or even the Axiom of Replacement, should ever be needed in scientific applications. However, remarkable recent investigations of Harvey Friedman's[18] suggest that the idea is not so outlandish after all.

---

[18] For a survey of recent results, see Harrington *et al.* [1985].

Since Gödel's incompleteness theorems, it has been known that mathematical theories incorporating strong existence axioms can have as deductive consequences sentences frameable in a lower-level mathematical language (e.g. the lowest level usually considered, that of the natural numbers) which are undecidable at that level (i.e. in a given theory appropriate to that level, say, PA^1). Thus, for example, if PA^1 is consistent, the number-theoretic assertion of its own consistency, $Con(PA^1)$, is undecidable in PA^1, but in any set theory T in which one can prove the existence of a model of the natural numbers one can also deduce $Con(PA^1)$. In general, systems involving objects of higher type have consequences "lower down" which may be new, that is, not demonstrable (or refutable) in the natural formalism of the lower level. Of course, one can always "fix" the lower-level theory by adding the desired consequence as an axiom, but (a) new undecidables will emerge, decidable at the richer level, and (b) such patched-up theories may be just that, *ad hoc* extensions not founded on any intuitively reasonable principles.

Because of this phenomenon of *non-conservativeness*[19] of richer mathematical theories with respect to lower-level theories, there arises the prospect of justifying the richer theories indirectly in virtue of their power to decide questions at the lower (more "observational") level that otherwise would remain undecided (except, perhaps, in *ad hoc* extensions). It was just this prospect that led Gödel to some of his well-known speculations concerning the possible justification of strong axioms of infinity for set theory.[20] One drawback to the programme, however, has been that, typically, the examples of undecidables (at the lower level) which get decided higher up were of a rather esoteric sort, statements that had little intuitive mathematical content and that depended on a great deal of coding in order to have any intuitive content at all. (Gödel sentences and statements of consistency were the stock-in-trade examples.) For some time, mathematical logicians have sought better examples, examples of undecidables with rather direct, mathematical significance, pertaining to problems that a mathematician might well pose in the course of mathematical—as oppsoed to metamathematical—work. A major result along these lines was that of Paris–Harrington [1977], in

---

[19] A theory T_2 is *conservative over* T_1 iff every sentence S in the language of T_1 that is provable in T_2 is also provable in T_1. If, in addition, every theorem of T_1 is a theorem of T_2, T_2 is called a *conservative extension* of T_1.
[20] See Gödel [1947].

which a number-theoretic statement of (a finite version of) Ramsey's partition theorem was shown to be unprovable in Peano arithmetic.

A whole series of remarkable results along similar lines has been obtained by Friedman [1981], in which the lower-level theory is already some axiomatic set theory (e.g. some natural modification of Z or ZF) and the higher-level theory is a richer set theory (e.g. ZFC or ZFC + a large cardinal axiom, etc.). But the statement in question is not an "esoteric" set-theoretic statement, but a statement pertaining to functions of a sort encountered in "normal" mathematics "lower down" (e.g. Borel functions from $\mathbb{R}^\omega \to \mathbb{R}$).[21]

Interesting, you may say, but what has all this got to do with applied mathematics? Perhaps nothing. However, the prospect is raised that among the "natural mathematical undecidables" could be found statements actually encountered in physics or some other empirical science. One can imagine that some sentence, S, unprovable in Z (or Z^+ with urelements), but statable even at the level of real numbers or functions on the reals, could be invoked in explaining some observable phenomenon. It might be that the existence of gaps in some sort of periodic table—expected, say, as an empirical regularity rather than on the basis of a rigorous mathematical symmetry argument—could best be explained as the reflection of an unsuspected mathematical phenomenon, say the inexistence of a certain type of function. (Perhaps tremendous resources could be saved—it would no longer be thought necessary to build the huge apparatus designed to find "particles" filling the gaps.) Yet to establish the mathematical fact, S, perhaps it would suffice to allow

---

[21] For a survey of results, see Nerode and Harrington, in Harrington *et al.* [1985], pp. 1–10. An example of an intuitive mathematical statement that requires uncountable iterations of the power set operation to prove (specifically, that is provable in ZFC but not in ZC) is the statement, "Every symmetric Borel subset of the unit square contains or is disjoint from the graph of a Borel function".

A series of results is based on analysing Cantor's diagonal argument that the unit interval I is uncountable. That argument produces a Borel diagonalization function F: $I^\omega \to I$ such that no $F(y)$ is a coordinate of y. But Cantor's F depends on the order in which the coordinates of y are given. If F is required to be *invariant* in the sense that $F(y) = F(y')$ whenever y and y' have the same coordinates, then the statement "If $F: I^\omega \to I$ is invariant, then some $F(x)$ is a coordinate of x", is provable in ZC but not in ZFC with the power set axiom deleted. Various modifications of this idea lead to examples of similar statements that are provable in ZFC + a large cardinal axiom, but not in ZFC. And there are even examples of statements of this sort which are provable in ZFC + "there is a measurable cardinal" but not in ZFC + "there is a Ramsey cardinal"! Thus, even at the level of "large, large cardinals" (incompatible with the Axiom of Constructibility, as Ramsey cardinals already are), it is necessary to go even further to prove certain "natural mathematical statements".

uncountable iterations of the power set operation (Replacement) or to assume the existence of a large cardinal. Doesn't the mere theoretical possibility of such a situation show that a "once and for all" framework for applied mathematics—as suggested in a scheme such as (3.3)—is fundamentally misguided?

This depends, I think, on how that framework is understood. As a scheme simply for *expressing* the statements of applied mathematics, something like (3.3) may well be defensible. But it should not therefore be inferred that such a framework demarcates any limit to "justification of mathematical axioms from below". Once we recognize (i) that justification comes in degrees, and (ii) that an axiom may receive justification indirectly in virtue of what can be *proved* using the axiom, even though a categorical framework is employed lower down—i.e. a framework which has only one model up to isomorphism and which, therefore, "semantically decides" everything formulable in the language—then there simply need be no such limit.

One objection to this line of reasoning should be considered. If the desired mathematical statement, S, of our highly speculative example is stable in the lower-level language, and if the (costly) empirical facts "confirm it", why not simply add it directly to the lower-level theory (e.g. Z^+)? The result would be somewhat analogous to adding the Axiom of Choice to ZF^2. Semantically, the addition is redundant (or else the result is inconsistent in the sense of "unsatisfiable in a full model"), since all full models of ZF^2 already agree on the Axiom of Choice (and, we assume, they all say "yes!"—we allow models to speak in various tones of voice!). Yet, proof-theoretically, the addition is significant, even at the level of second-order systems.[22] If this course were followed, there would seem to be no role left for the higher-level set-theoretic axioms to play.

The response to this, I think, is twofold. First, adopting S outright as an axiom may have nothing to recommend it other than its fit with the empirical observations. It may have no independent mathematical motivation. As a result, the move could count as *ad hoc* and not preferable to some alternative, such as a modification in the physical theory itself. In comparison, a strong set-theoretic principle, such as Replacement, may have all kinds of independent mathematical motivation. Considerations such as those encountered above (Chapter 2, §4) would be relevant; and many other consequences

---

[22] NB. The independence proofs of Cohen [1966] can be adapted to second-order axiomatizations of set theory. See Chuaqui [1972].

internal to mathematics could be adduced. (For example, in the presence of Choice and Replacement, cardinal arithmetic becomes more unified, in that cardinals can be taken as initial ordinals, all sets have comparable cardinalities, etc.) If the justification of mathematical axioms is a multifaceted affair (as we believe it to be), with partial justification stemming from various sources, then we may be comparing a situation (adopting S as an axiom, i.e. modal existence of a $Z^+ + S$ structure) in which the *only* source is empirical with one (adopting ZF, i.e. modal existence of a ZF structure) in which there is already independent partial justification. Thus, the two situations need by no means be equivalent.

The second part of the response is really an extension of the first. Even if S were to have independent mathematical motivation as a new axiom, S might be only one among a large number of "physically significant undecidables". Suppose it happened that each of these could be proven from the *same* higher set-theoretic principle, but in the absence of this they were mutually independent. They could, moreover, "sound" as intuitively different as you please. Surely, the higher principle here would be playing a unifying role of the very sort that guides theory construction in general. Rather than rest with the highly disparate result of adding each undecidable (perhaps there are even infinitely many) as a separate axiom, surely adoption of the higher set-theoretic principle would seem the reasonable course.

In sum, we are lead to the view that a global scheme for applied mathematics, such as (3.3), can at best be claimed to be expressively adequate. The thrust behind our speculations concerning Gödel's and Friedman's programme concerning justification of higher axioms from below should be taken seriously. Discoveries at the level of empirical science could indeed transmit a sort of justification upward, well beyond the limit of expressive adequacy. And, indeed, it does seem misguided to seek any a priori limit to this process.

§4. "Metaphysical Realist" Commitments? "Synthetic Determination" Relations

Let us now return to the issue raised above concerning the proviso of "no-disturbance" built into our understanding of the applied mathematical counterfactuals. When entertaining any such counterfactual, we must of course assume that the actual situation in question is "held constant". In invoking a suitable mathematical structure, about whose actual existence we wish to remain neutral, we are of course not contemplating any "change" in the actual non-mathematical situation, apart from what would logically follow from the assumption of the mathematical structure (such as the relations that would arise in virtue of our background logical comprehension principles). As we might naïvely put it: "the real, material situation is supposed to remain exactly as it is."[23]

Now, so long as we are prepared to rest with such a blanket formulation, we can employ the counterfactuals in the form already illustrated above. If called upon to say how it is we know that the consequents of the conditionals carry information about the actual world or system in question, and not some different world, our answer must be, "we stipulated at the outset that all matters of material fact are held fixed; that is what we understand by this counterfactual idiom. What's the problem?"[24]

Philosophy is in the business of finding problems, and then wrangling over whether the problem is merely a *Scheinproblem* (or perhaps a problem of a worse kind (replace the '*n*' in '*Schein*' by 'ss'!)). (This procedure can then be iterated, guaranteeing an endless supply of problems. At limits, λ, take the problem to be, "What is the λth problem?" This yields transfinitely many problems!) In the present case, the problem (if it is one) seems to be that, if called upon to be more specific about the material matters of fact that are assumed fixed, we very readily find ourselves invoking applied

[23] It would be better to say "real non-mathematical situation", since presumably the problems here arise independently of commitment to a materialist ontology. Dualists, phenomenalists, *et al.*, may seek to apply mathematics in various ways, and they too would need to invoke an assumption of non-interference or fixity of the non-mathematical world recognized. Here and below, this should be understood, though for brevity we shall sometimes speak of "material conditions", etc.

[24] Something like such a stipulation likes behind Horgan's [1987] reasonable solution to a related problem raised by Hale and Resnik [1987] concerning Horgan's use of counterfactuals [1984] to "nominalize" applied mathematics. The objection was that ordinary patterns of explanation of mathematics-free statements, O (e.g. "observation statements"), would be upset under a counterfactual interpretation, because in the modalized theory one would deduce, not O, but something of the form $\square(A \supset O)$ (where A is an appropriate antecedent involving mathematical axioms). Horgan's solution is, in such cases, to add as an axiom,

$$(\square(A \supset O)) \supset O,$$

on the grounds that, if O were to hold under the conditions envisioned in A, then it holds in fact. This is reasonable because in entertaining A, we entertain no alteration of non-mathematical matters of fact.

mathematical descriptions. We may say that the relative positions and velocities of these masses are to remain fixed, or that these physical fields are to retain their actual values at every point, and so forth. Of course, we don't confine ourselves to such descriptions. There is much we can say about our material surroundings in qualitative terms, without apparent reference to mathematical objects. But the point is that we are hard-pressed to specify in any kind of precise detail "physical or material conditions" without appealing to our best theoretical descriptions, and in these apparent mathematical reference is rife. If we attempt to eliminate such reference along the lines of our modal conditionals, we need to stipulate that material conditions are to be held fixed in understanding *these*. But if it is insisted that we spell this out in detail, we find ourselves employing non-hypothetical applied mathematics ...

There are, broadly speaking, two lines of response to this circularity problem that can be pursued. Each is associated with a philosophical position on questions of (non-mathemetical) realism. The first line *rejects the problem*: it rejects the demand for any detailed specification of "facts to be held fixed" in the antecedents of the counterfactuals, resting instead with the general formulations already given. For reasons that will emerge in a moment, this view seems to be committed to a certain kind of "realist" view of nature, a realism that has been challenged lately in the philosophical literature.[25] The second line of response accepts the need for greater specificity in spelling out "the facts to be held fixed", but attempts to avoid the circularity just noted by framing the relevant conditions in "synthetic, relational terms", that is, in a vocabulary involving solely relations among non-mathematical objects and not containing any apparent reference to abstract mathematical objects (numbers, functions, sets, etc.). (Here and below, such vocabulary is called "synthetic" by analogy with synthetic geometry, which is carried out solely in terms of relations among geometric objects (e.g. 'betweenness', 'congruence', etc.).) The aim of this approach is to allow for neutrality on the "metaphysical commitments" associated with the first approach. At the same time, it bears an interesting resemblance to certain recent efforts to sustain a kind of "nominalism" by attempting to demonstrate that theories framed entirely in terms of relations on a non-mathematical domain (e.g. space-time) can capture

---

[25] See especially recent writings of Goodman [1978] and Putnam [1981].

the content—for scientific purposes—of standard mathematical-physical theories.[26] Let us examine these two approaches in turn.

The first approach is committed to a realist view of nature that might be put roughly as follows: in science, we investigate a unique material reality, existing objectively and independent of our minds, our language, conceptual scheme, and so forth.[27] Of course, this world can be described in many diverse ways, some of which may even be "incommensurate" with others (in various senses); and, of course, our descriptions, conceptualizations, and so forth, are "mind-dependent". When it comes to mathematics, however, we need not regard its abstract structures as literally part of the actual world. It is sufficient that they be conceptually possible. In applying mathematics to the actual world, we appeal to mathematical structures as a way of carrying information about that world. In our own language, we may well not have any alternative way of expressing this information; mathematical language may be indispensable for any precise, detailed description, especially one which we can use in a theory for purposes of prediction and explanation. Nevertheless, this indispensability reflects our own language, perhaps our own capacities. It doesn't mean, for example, that abstract mathematical objects literally participate with the non-mathematical in making up material reality. When, for instance, we describe the relative positions of a bunch of particles (at a time, or over an interval of time), we find it convenient—perhaps for certain purposes, indispensable—to invoke numerical coordinates (e.g. to assign a triple or quadruple of real numbers). We think of position as a function from material particles to (triples or quadruples) of real numbers. But, quite independently of this mathematical function, in fact the particles have their relative locations or spatio-temporal distribution. These material relationships are part of material reality as much as the particles themselves: the system consists of particles-in-a-particular-configuration. Mathematics is convenient (perhaps essential) in describing that configuration in detail, but the configuration is "already there" prior to

[26] I refer here to work of Field [1980] and Burgess [1984] to which we will return below.

[27] As the term "independent" is quite ambiguous, this sort of statement (or its negation, favoured by various "anti-realisms") must be articulated with some care. In fact, it is non-trivial to find notions of "mind-independence" that can be used even to differentiate "realist" and "instrumentalist" positions from one another. For an attempt in this direction and a discussion of some of the problems, see Hellman [1983].

the mathematical description, and independent of any (quite arbitrary) structure for real numbers that may be invoked in such a description. Similarly for physical fields. We describe them as, say, vector-valued functions on a spatio-temporal domain, but we should not therefore identify the field with a mathematical representation of it (i.e. with an information-bearing mathematical function—the *linguistic* description is, of course, not in question here). Fields have physical effects; their mathematical counterparts carry information as to what those effects will or would be under various conditions, but, as abstract objects, they themselves don't literally *have* the effects. A sofa, too, presumably, could be given a mathematical description, via a density function (on a region of space-time, say). But we do not literally identify the sofa and the density function.

On this view, material reality consists of a great diversity of objects interrelated in a rich variety of ways, only some of which are gotten at at all in our schematic, human-based languages and theories. Obviously, we cannot say anything about this reality without introducing our own predicates, concepts, and so forth. But this doesn't vitiate the idea of material reality; it just limits the usefulness of that idea. But limited usefulness is not complete uselessness, and, in any case, uselessness is not incoherence. And, in applied modal mathematics, we see that the idea does have some use: as a very general and schematic device for helping us say what we are doing when we describe the world mathematically.[28]

As reasonable as this natural realism may seem, it has its challengers. It has been argued that it is incoherent to speak of a unique "way that the world is" or even of a unique "world": the world is as many ways as it can be truly described; or, better, we should not even speak of "the world" but only of "world-versions", "world descriptions", "world-portrayals", and so on.[29] Now the above outlined realist perspective certainly avoids any commitment to a unique correct description of material reality. Still, it seems committed to material properties and relations as well as objects, whether describable or expressible in human language or not.[30] And this is seen as

---

[28] This may be considered a partial response to Goodman [1978] which emphasizes the emptiness of reference to a unique world underlying our own constructions. It is not by any means suggested here that the role of such reference in applied modal mathematics is its only or principal use.

[29] Goodman [1972] and [1978].

[30] The above sketch of a realist position has interesting affinities with Putnam's "On Properties" (in [1970]), reflecting an earlier realist perspective.

"metaphysical" in a pejorative sense. The realist could respond that the position is nothing more than an objective view of science; that material reality is what science investigates; and that the commitment should be called "hypo-physical" rather than "metaphysical" ('hypo' for "under" or "supporting"). In any case, realist commitments there are, and those who seek to avoid them will want to know whether they can succeed within a modal-structural framework for applied mathematics.

This brings us to the second approach. Here one attempts to provide a specification of relational (synthetic) predicates on the (non-mathematical) domain of mathematical application which can be utilized in the antecedents of the counterfactuals to achieve the effect of "fixing the actual material situation". What is required of such a vocabulary?

First, we must suppose that the synthetic predicates, say of a finite list $R_1 \ldots R_n$, are "ontologically adequate" in the sense that every non-mathematical object of interest in the mathematical application in question falls within the actual extensions of some predicate on the list. But more than this is clearly necessary. To see what is involved: suppose that mutually conflicting mathematical descriptions, $d_1, d_2, \ldots$, of an actual situation were compatible with a given description of that situation in terms of the R_i; then we could not truly assert any of the respective counterfactuals containing the d_i (or their counterparts generated by the translation pattern) as consequents. In effect, one would not have "fixed enough" to know *which* applied mathematical statements applied to the actual situation. Suppose that the R_i were insufficient to determine, say, the mass of a certain object o. Then the modal conditional associated with the ordinary statement "o has mass r" could not be affirmed: it would not be the case that, necessarily, were there a suitable mathematical structure respecting the R predicates (i.e. in which the R predicates had their actual extensions), there would be a mass-representing function assigning o the value r. For, by hypothesis, there could be mathematical structures respecting the R predicates and different mass-representing functions associated with those structures assigning o different values, each compatible with the R description. (The different functions would still be "mass-representing" in giving "actually measured and theoretically predicted values of mass" in different *physically* possible worlds. The conditions on magnitude-representation given above should be understood as "world relative" in this sense; different choices of "the actual world" are possible, in that a physical theory can also be "applied" counterfactually, i.e. under assumptions of non-actual initial conditions, forces, etc. But of course, any such physically possible situations count at logically possible as well, and so come within the scope of the '$\Box$' of the applied mathematical counterfactuals.)

If, however, one were convinced that the R predicates sufficed to "determine uniquely" a mathematical-physical description of the situation (taking account, of course, the fixing of all arbitrary elements of such a description, such as units of measurement, choices of guage for fields, choices of purely mathematical constructions, and so forth), then one could employ the list in stating mathematical counterfactuals explicitly, along the following lines:

$$\Box \forall X \forall f \left\{ (\wedge Z^+)^X \binom{\epsilon}{f} \,\&\, \forall x (x \text{ an urelement of } X \equiv \right.$$
$$@ \vee_{i=1}^n R_i(\ldots x \ldots) \,\&\,$$
$$\left. \wedge_{i=1}^n \forall x_1 \ldots \forall x_{k_i}[R_i^{k_i}(x_1 \ldots x_{k_i}) \equiv @R_i^{k_i}(x_1 \ldots x_{k_i})] \supset \ldots \right\}, \quad (3.4)$$

where we have employed the global scheme of the previous section. The second conjunct says that the urelements of the Z^+ model are just those items that are actually ($@$) relata of the R_i (where $R_i(\ldots x \ldots)$ is abbreviatory of the obvious long-winded statement of this; and the third conjunct says that each of the R_i behaves in the model just as it actually does.

Provided we were convinced of the adequacy of the R_i (in the above sense of ontological exhaustiveness and as a determination base for mathematical-physical description), the pattern of (3.4) could be employed apart from the realist commitments of the first approach. For here, we do no more than employ predicates with modal operators. We say such things as "these things are actually F" or "the segment p_1–p_2 is actually congruent to the segment q_1–q_2" (to take a simple statement typical of synthetic geometry); we never need to speak directly of properties or relations in reality. Questions of reference between linguistic predicates and material properties and relations can be postponed, sidestepped, or even rejected as "meaningless". In this sense, the scheme (3.4) is "metaphysically (or hypo-physically) neutral".

There is a way in which (3.4) can be broadened without (substantially) violating this neutrality: that is to allow the use of a semantic relation of "application" or "true-of" between (relational) predicates and first-order objects (particles, fields, space-time regions, whatever). That is, we allow clauses in the antecedent of (3.4) such as

'heavier than' is true of $\langle x, y \rangle \equiv$ @ 'heavier than' is true of $\langle x, y \rangle$

(where it must be assumed that some concrete specification of ordered pairs is available).[31] In that case, the restriction to finite bases in (3.4) can be lifted; provided we have some way of specifying a class of 'non-mathematical' predicates—say, by reference to syntactic rules for constructing them—the second and third conjuncts of (3.4) could be rewritten by quantifying over such predicates and stating the desired conditions in terms of 'true-of' as just illustrated. The "realist commitments" of such a framework, whatever they may be precisely, are surely more modest than those of a full-fledged (material) property realism. Nothing more than the semantic relation of ordinary reference to first-order objects is presupposed.

Once this latter move is made, the possibilities of adequate synthetic bases become quite enormous. For now we can even allow predicates of the form, "x has mass c" for particular constants, c (c can be an arbitrary rational, or any real for which a notation can be described). We could employ predicates of the form, "x and y are separated by distance bearing ratio r to standard length l", where 'r' is a rational constant and 'l' is a constant designating a preselected fixed standard (e.g. a well-isolated metre stick). Such predicates do not involve quantification over numbers; and our understanding, of them—sufficient for their use in a scheme such as (3.4)—can perhaps be explained operationally, without quantifying over numbers or other mathematical objects. By invoking sufficiently many predicates of this sort, one may hope to supply the required "fixation of the material facts" without circularity, and without strong hypophysical commitments.

But how can we become convinced that indeed a proposed "synthetic basis"—finite or infinite—is adequate? Adequacy, recall,

---

[31] If there is a finite upper bound on the number of places required in the predicates of a synthetic determination base, then a synthetic apparatus of ordered tuples can be dispensed with in favour of finitely many $k + 1$-ary "true-of" predicates, of the form

'P^k' is-true-of $x_1, x_2, \ldots, x_k$,

in which the order of places is built into the predicate. The presumption would be that finitely many such designation relations are learnable one by one.

involves two parts: ontological adequacy and adequacy as a "determination base" for the applied mathematical descriptions in question. The former is relatively unproblematic, for the particular context of application usually involves a given domain of material objects to which a piece of mathematics is to be applied, and it suffices to cover this domain with the synthetic predicates. "Determination" is more problematic, and it must be made more precise to be evaluated.

One natural way of proceeding is to invoke models of an overall theory T' including both the vocabulary of the applied mathematical theory, T, and the proposed synthetic vocabulary, S. Such a theory links the two vocabularies, and may be assumed to be an extension of T, as well. T, we may assume, specifies, up to isomorphism, a particular type of mathematical structure (e.g. it might contain Z^+). Then we may explicate determination along the following lines:

> Let α be the class of (mathematically) standard models of T', and let V denote the full vocabulary of T': then S determines V in α iff for any two models m and m' in α, and any bijection ϕ between their domains, if ϕ is an S isomorphism, it is also a V isomorphism.

The last clause means that if ϕ preserves the synthetic vocabulary, it also preserves the rest of the vocabulary of the total theory T', including all relations between the non-mathematical part of the domain and the mathematical part.[32]

[32] Determination principles of roughly this general form were introduced in Hellman and Thompson [1975] as a means of explicating a kind of physicalism not committed to strong claims of reducibility (definability of determined by determining vocabulary). Related principles, known as "supervenience principles", have been developed and applied in a variety of contexts. See especially Kim [1984]. For a survey, see Teller [1983]. See also Post [1987] and Hellman [1985] and references therein. The present sort of determination principle differs from the usual in that, here, one is expanding upon a well-worked-out theory for the "higher" level, rather than for the "lower".

Note that the restriction to mathematically standard (full) models obviates an automatic collapse to explicit definability (of determined by determining vocabulary) via the Beth definability theorem. (Cf. e.g. Shoenfield [1967] for a precise statement and proof of the theorem; for discussion, see Hellman and Thompson [1975].)

Note further that ("vertical") determination claims are compatible with ("horizontal") indeterminism in temporal evolution: they merely imply that whatever temporal branching is possible in the higher (determined) level must already be reflected in corresponding branching at the lower (determining) level.

Finally, note that it is not required that the synthetic vocabulary here be "observational" in any restrictive sense. It may, in its own right, count as highly "theoretical"; and, if the device of 'true-of' semantic predicates is allowed, theoretical vocabulary of

The idea now would be that for suitably developed T′ and S, a determination claim of this sort could be supported, either inductively, by examining a sufficient variety of particular systems to which T′ applied (a procedure that becomes somewhat unwieldly when T is global cosmology and you are not God); or deductively by a mathematical argument. A merely suggestive (and intentionally rather contrived) example of how the latter might come about, follows. It will serve us below.

Let the original applied mathematical theory T be one involving a real scalar field ϕ defined on a region of space-time, and satisfying an axiom of continuity. (T is also assumed to contain axioms for the real numbers.) Suppose that ϕ is thought to represent a physical magnitude M, and further suppose that for each pair of rationals, q and ε, we introduce a predicate, $M_{q,\varepsilon}(x)$, intuitively understood as saying that the value of M at point x is within ε of q. (x is assumed to have real coordinates in some given system; ε could be canonically chosen as of the form 2^{-n}, for positive integral n.) (The use of bold face for $\mathbf{q}$ and $\boldsymbol{\varepsilon}$ is to indicate that these are non-quantifiable parts of the new predicates; italicized letters behave as ordinary quantifiable variables. Clearly, effective rules could be given for the construction of these M predicates.) Now let T′ be the result of adding to T countably many axioms, one for each q and ε, of the following form:

$\forall x[x$ a point in the domain of ϕ with rational coordinates $\supset$

$$(M_{q,\varepsilon}(x) \equiv |\phi(x) - q| < \varepsilon)].$$

Now, appealing to the mathematical facts (i) that $\phi(x)$ is the limit of $\langle\phi(q_i)\rangle$ for Cauchy sequences of rationals $\langle q_i\rangle$ with limit x, and (ii) that ϕ is uniquely determined by its values at rational points, it follows that if two mathematically standard models of T′ agree on the M predicates, they also agree on 'ϕ'. Thus, the M predicates determine ϕ (relative to standard models of T′). And these predicates could qualify as "synthetic" since they could be understood as wholes, based on measurement operations (including symbolic manipulations), not as involving any genuine reference to rationals. Similar tricks will clearly work for continuous vector fields; and the

the original applied mathematical theory may be adapted to the synthetic level by use of countably many instances involving rational values, as suggested above (cf. the example that follows below).

requirement of continuity could be relaxed to allow for countably many discontinuities.

Before considering the import of this sort of phenomenon, it would be well to forestall an objection: A determination claim along the above lines employs model theory, and so is platonistic as formulated. How can the modalist make use of such a claim, much less prove it in particular cases? The answer is twofold: first of all, as formulated, such a claim can be understood abstractly; there is no *use* of non-mathematical vocabulary; it is a purely model-theoretic claim concerning the interconnectedness of parts of models. As such, it could be modalized along with model theory generally as part of set theory (along the lines of Chapter 2). Secondly, however, the claim may be read as a step in a "*reductio*" argument from within platonism, in the manner of ontological reductions generally.[33] From the assumptions of platonistic mathematics, if one can arrive at suitable determination results, one can then see that applied mathematics can be carried out modally, without circularity; hence, those assumptions can be dropped (at least, in so far as applied mathematics is concerned).[34]

Suppose, then, that a determination claim of the above variety—to the effect that a specifiable "synthetic vocabulary" uniquely determines analytical mathematical description within a given scientific theory, T—can be supported. Does this show that the analytic theory T can be eliminated in favour of a synthetic alternative? No, this does not follow. The key point here is that what does the determining is a class of predicates, linked to mathematical predicates within a theory. (The links may be thought of as provided by an extension T' of the original applied mathematical theory T, if T does

---

[33] Cf. Quine [1969].

[34] This ontological *reductio* pattern is the backbone structure of Field [1980]: the representation theorem is a theorem of model theory, accessible to the platonist, but not to the nominalist (at least, not without further argument, which was not provided in Field [1980]). It should be pointed out, however, that such *reductio* reasoning is problematic in a certain sense: how can it convince one who starts out with the assumption that platonist reference to models, etc., is unintelligible? For one in that position, the *reductio* argument itself should be unintelligible. But how, then, could a "nominalist scientist" become convinced (by Field-theoretic reasoning) that the standard platonist reasoning (which, of course, the nominalist scientist employs every day) is "just a short-cut"? One possible answer here involves distinguishing between a non-constructive nominalist position—which holds platonist ontological commitments to be false, but not unintelligible—and a stricter constructive-nominalist position, which does hold those commitments to be unintelligible. Then perhaps we should say that only the latter is barred from following Field-theoretic reasoning.

not already supply them.) Presumably this theory (T′)—and, more importantly, the original T, which T′ extends, is well formulated in various ways. (For example, it may be finitely axiomatizable; presumably it is at least recursively axiomatizable. And it may be otherwise "attractive".) But there is no guarantee that a correspondingly *attractive theory* is formulable in just the synthetic vocabulary. Indeed the set of all deductive consequences of T′ in the synthetic vocabulary will presumably be recursively enumerable; but, as considerations of Craigian replacements in other contexts have brought out, this is certainly not sufficient for "good systematization", not to mention other aspects of "attractiveness".[35] The above example illustrates this: there is no *hint* of an attractive theory involving just the M predicates. (For instance, there is no direct way even to formulate an axiom of continuity using the M predicates, much less laws interrelating the field ϕ with other quantities that might readily be stated in extensions of the theory T.) Thus, even if a synthetic determination claim can be supported or proved, this by no means shows that the original applied mathematical theory is really dispensable for scientific purposes. For that, much more would have to be demonstrated.

Now there is an alternative programme in the foundations of applied mathematics—that of Field [1980]—which seeks just such a demonstration. It is worth examining as a very interesting approach in its own right. Moreover, comparison with the present approach is instructive. For while there are some apparent similarities between the two approaches, there are also some fundamental differences which deserve to be highlighted.

§ 5. A Role for Representation Theorems?

In his book, *Science without Numbers*, Field sought to extend work in the foundations of geometry to mathematical physics with the aim of showing that ordinary applied mathematical reasoning can in principle be replaced by reasoning entirely on a synthetic level, that

---

[35] Craig's observation—that any recursively enumerable theory has a recursive axiomatization—is in Craig [1953]. For a discussion and critique of its use in philosophy of science, see Putnam [1965]. It should be noted that Field [1980], p. 47, explicitly eschews a Craigian replacement of platonistic applied mathematics (i.e. a recursive axiomatization of its nominalistic consequences) as even a candidate for a nominalization, on grounds of its obvious unattractiveness as a theory.

is, in terms of relations on a non-mathematical domain, such as space-time. Much as Hilbert's work on geometry [1971] showed that reasoning concerning metrical relations, carried out in terms of real numbers and real-valued functions, could be replaced by reasoning entirely on the level of intrinsic geometrical relations (such as relations of *betweenness* and *congruence*), Field sought to show that a similar replacement could be effected for reasoning involving scalar and vector fields pertaining to physical magnitudes. A key step in the programme is proof of a "representation theorem", in the style of measurement theory,[36] which asserts a strong correspondence between models of the synthetic replacing theory and those of the original applied mathematical theory to be replaced.[37] In outline, the form of such a theorem is as follows:

An original applied mathematical theory (e.g. a branch of physics), T, involving, say, a scalar field ψ (and other similar objects, which we shall suppress for brevity), is given and determines a class of models of a certain form, say, $((M, d), \psi)$, where d is a metric or distance function on a manifold, M. A replacing theory T_{syn} is constructed in a privileged synthetic language, which might include, say, a relation of segment congruence, Seg-Cong(x, y, z, w), meaning intuitively that the segment from (space-time) point x to y is congruent to the segment from z to w, i.e. as measured via d; and further a relation of *scale-betweenness*, Scale-Bet(x, y, z), holding intuitively when the value of ψ at y is (inclusively) between those at x and z; etc. The final preliminary is that T_{syn} contain machinery essentially equivalent to second-order monadic predicate logic: a standard interpretation of T_{syn} will be *full*, in the sense that the range of the second-order quantifiers will be all subsets (or regions) of the synthetic geometrical domain. The representation theorem then states:

> For every standard interpretation $(X, Seg\text{-}Cong_X, \text{Scale-}Bet_X, \ldots)$ which is a model of T_{syn} there exists a model $((M, d), \psi)$ of T and a homomorphism between the two models, i.e. a bijection $\phi: X \to M$ such that $\forall p, q, r, s$ of X:
>
> Seg-Cong$_X(p, q, r, s)$ iff $d[\phi(p), \phi(q)] = d[\phi(r), \phi(s)]$, and
>
> Scale-Bet$_X(p, q, r)$ iff $\psi(\phi(p)) \leq \psi(\phi(q)) \leq \psi(\phi(r))$,

[36] As in Krantz *et al.* [1971].

[37] For details on representation theorems, see Field [1980], Ch. 7; see also Malament [1982]. In our sketch below we follow the notation of Malament's review (which concentrated on the elegant example of the Klein–Gordon field).

and similarly for other relations of the theory; moreover, the model of T and the homomorphism ϕ are *essentially unique*, i.e. unique up to a transformation within a given class of transformations reflecting arbitrary choices such as units of measurement.

Given such a theorem, Field then attempted to argue that any synthetically statable sentence deducible from T—i.e. making full use of platonistic mathematics—could in principle already be deduced within T_{syn}, i.e. confining oneself to the (allegedly nominalistically acceptable) synthetic vocabulary. Now, it is clear, given such a representation theorem, that the model-theoretic analogue of this statement holds, that is, that synthetic logical consequences of T are also logical consequences of T_{syn} (where this is spelled out in the usual way in terms of full second-order models). However, the proof-theoretic conservativeness is not automatically implied, and, in fact, in a large class of cases it can be proved not to hold. We will return to this point in a moment.[38]

Now consider the relationship between a Field-style representation theorem and a claim of synthetic determination. Clearly they are similar in spirit: given the uniqueness clause of the representation theorem, once a synthetic model is given (once "the synthetic facts are fixed"), there is determined an essentially unique applied mathematical model ("the applied mathematical facts are fixed" also). Thus proving a representation theorem is intuitively quite close to establishing a corresponding synthetic determination claim. But the connection is not perfectly tight, and especially when we attempt to argue the converse direction, we run into difficulties. For, as we have already seen, there is no guarantee of a "respectable" theory at the synthetic level. Although this isn't a formal condition of a representation theorem, it is essential if such a theorem is to have the force it is

[38] The proof-theoretic conservativeness claim is naturally read as the relevant claim that Field [1980] sought to establish. However, Field [1985] emphasizes that it is the semantic conservativeness of mathematical physics over synthetic physics that should be the aim of the programme. (He there writes, "What I should have said is that mathematics is useful because it is often easier to see that a nominalistic claim follows from a nominalistic theory plus mathematics than to see that it follows from the nominalistic theory alone" (p. 241).) As I will have occasion to remark below, I do not think that this aim really comes to grips with the issue of indispensability of higher mathematics—indispensability in its principal role of proving theorems. Ironically, this (so far as we know, indispensable) role appears to be conceded by Field [1985] in his very argument that semantic conservativeness claims can be put to use despite the lack of a complete proof procedure (see Field [1985], p. 252).

supposed to have concerning dispensability of analytical mathematics. (Other difficulties with the converse inference in question concern the domains of models; these are of a more technical nature and could probably be patched up.) This brings out a critical difference between the two approaches: On the Field programme, it is essential to find acceptable *substitute theories* in which mathematical physics can actually be carried out. This goes well beyond the present ms reconstruction, which merely needs assurance at a certain point (and under certain requirements of "metaphysical neutrality") that its translates are determinate. For this, it is enough to be confident that a synthetic vocabulary is *descriptively adequate* in a limited sense; it is by no means necessary to argue in addition that a good theory in this vocabulary is forthcoming over which the applied mathematics is conservative. In sum, the Field programme seeks to show that mathematically formulated theories are in principle dispensable for scientific purposes. Not only does the present approach seek no such goal; we have already suggested that non-conservativeness results in higher mathematics may turn out to have physically significant corollaries, and that no line can in principle be drawn beyond which—proof-theoretically at any rate—rich mathematical theories become scientifically irrelevant.

Now, while it should be clear that, on an abstract level, determination claims are weaker than representation theorems, it might still be argued that the latter are the best access we have—mathematically—to the former, that, in practice, a representation theorem should still be sought by the modal structuralist. But then—the argument might run—once such a theorem were proved, it would follow that the platonistic mathematics was dispensable after all, and there wouldn't even be a need for a modal-structural reconstruction (from a standpoint seeking alternatives to standard platonism). The answer to this is twofold. First, as the example of § 4 suggests, there are ways of bypassing representation theorems, at least in a wide class of cases. But, second, this line of argument is in any case fundamentally flawed. For, as has already been mentioned, even a full-fledged representation theorem does not demonstrate deductive conservativeness of the original mathematical physical theory T over the synthetic substitute T_{syn}. As several authors have pointed out, when account is taken of the second-order machinery of T_{syn}, which is actually needed to establish Field's representation theorem (e.g. to prove the existence of a representing homomorphism between space-time and $\mathbb{R}^4$), it can be shown that, for example, standard number theory can be carried out within T_{syn} and Gödel sentences concerning provability in T_{syn} (which is an *axiomatic* second-order theory, to which the Gödel theorems apply) arise, such as assertions of T_{syn}'s consistency, etc.[39] Although not provable in T_{syn} (on pain of its inconsistency, in which case the representation theorem would be vacuous), such sentences, in the language of T_{syn}, *are* provable in suitably rich applied mathematical theories (in particular, in ZFC with urelements, which Field exploited in his presentation). And, although such examples may be *recherché* from the point of view of physics, the results of Paris–Harrington, Friedman, and others (cf. above, §3), suggest that it would be premature at best to suppose that all such violations of conservativeness are "of no physical significance".

At this point, one may be tempted to retreat to claims of "semantic conservativeness" of platonistic applied mathematics over synthetic theories: that is, although the former can prove synthetic statements that the latter cannot, still all synthetically stable semantic consequences of the former are indeed semantic consequences of the latter. In the cases discussed in the literature, this is true, and is a fairly trivial consequence of the definition of (second-order) semantic consequence. However, as I see it, this is quite beside the point. For one of the uses of higher mathematics is to *prove* theorems that may be stable lower down. Although the lower-level theory—if formulated in second-order logic—has a negation-complete set of semantic consequences, it provides no method for deciding what these are, nor even a method of enumerating them. Now, even though no recursively axiomatizable theory can prove all these second-order semantic consequences, deductively stronger mathematical theories provide more information about particular cases, and these may be scientifically important. If higher mathematics is indispensable in this sense—i.e. in the sense that it is needed to *prove* truths lower down of scientific importance—then surely that is indispensability

---

[39] See especially Shapiro [1983*b*] and Burgess [1984]. Field [1980] acknowledges the difficulty at pp. 104 ff., crediting Moschovakis and Burgess with the observation that Gödel sentences arise as counterexamples to the conservativeness of T over T_{syn}. Here we emphasize the point that increasingly powerful mathematical theories may well be relevant for proving physically meaningful synthetic assertions. The fact that, as axiomatized theories, these more powerful systems have their own undecidables, is, from the present perspective, if anything, a reason for thinking that we may always have to consider stronger and stronger theories.

enough for the purposes of the usual pro-platonist arguments. In no way is its force weakened by the quite different consideration of semantic conservativeness.[40]

Thus, proof of a Field-style representation theorem for a physical theory does not carry with it a general implication of dispensability of the higher mathematics employed within that theory. Such a theorem may, however, yield insight on the mechanism whereby mathematics can be successfully applied to a given non-mathematical domain (as, for example, a representing homomorphism from space-time to $\mathbb{R}^4$ can be claimed to provide). And, as already observed, such theorems can be of help in justifying a modal-structural reconstruction. But I see no general inference in this quarter that success in proving a representation theorem undermines the need for the reconstruction.

The need might be undermined in other ways, however, and attention should be called here to an alternative approach of Burgess's [1984], which does indeed yield deductive conservativeness results in certain cases of the sort that Field considered. In this approach, one begins with a preferred synthetic vocabulary and builds a theory T_{syn} designed to "re-express" in synthetic terms what a given original applied mathematical theory expresses about a non-mathematical domain (e.g. space-time, in the examples actually developed). To demonstrate that the "re-expression" is faithful, one appeals, not to representation theorems in the style of measurement theory, but to strictly first-order proof-theoretic considerations: namely one builds a conservative extension T_{ext} of T_{syn}, by direct logical constructions—introduction of explicit definitions, new constants, and equivalence relations permitting conservative use of abstraction operators—and then one shows that in T_{ext} one can actually derive the theorems of T.[41] The conservativeness of T_{ext}—in the relevant deductive sense—is a direct consequence of its methods of construction

---

[40] Other objections to appeals to semantic conservativeness have been raised, in particular that the nominalist should not be able to understand "semantic consequence" in the relevant (full, second-order) sense, since it involves quantification over abstract models (cf. Malament [1982]). Prima facie, this seems telling. However, by means of coding devices, a good deal of abstract model theory (for theories such as PA and RA) can be carried out "nominalistically" (cf. Ch. 1, §6), i.e. in monadic second-order systems, especially when interpreted over space-time.

Another point in the response to the "unintelligibility" objection would be that appeals to semantic conservativeness are part of the overall *reductio* argument, which makes free use of platonist constructions, as does the representation theorem itself. Our remarks above, n. 34, would then apply here as well.

[41] For details, see Burgess [1984].

(reflected in general facts about first-order logic, collected in Burgess [1984]). And, while it may appear that the machinery for a Field-style representation theorem is available, and hence (in light of Shapiro's observations in [1983]) that Gödel sentences and other witnesses to non-conservativeness ought to arise, the appearance is illusory. For, in fact, since everything is carried out in first-order, non-standard interpretations at the level of T_{syn} and T_{ext} are not ruled out, so that there is no way to prove the existence of a representing homomorphism (e.g. from space-time to $\mathbb{R}^4$), and no way to provide a biconditional link between, say, the ordinary mathematical statement of T_{syn}'s consistency, $\text{Con}_{T_{syn}}(\omega, 0)$ (which, indeed, T_{ext} may prove), and the statement which, standardly interpreted, expresses this in the language of T_{syn} itself.[42]

This direct logical approach appears promising as a means of providing synthetic alternatives to certain mathematical-physical theories. The method has been illustrated for classical field theories in which one has flat space-time as a background. It will be very interesting to see whether it can be extended to the curved space-time framework of General Relativity, and whether it can be adapted to non-classical (quantum mechanical) particle and field theories. In both cases, questions arise in the choice of synthetic primitives; and, especially in the case of quantum theories, whose models are already highly abstract (involving, in modern formulations, lattices of subspaces of an already abstract Hilbert space), it is by no means evident that interesting synthetic reformulations are possible.[43]

It should also be pointed out that, even in the case of flat space-time theories, the synthetic alternatives proposed in the literature (along the lines of both Field [1980] and Burgess [1984]) involve a commitment to space-time points as objects, and this is a commitment that is subject to a number of objections commonly levelled against the objects of platonist mathematics itself. It is unclear how

---

[42] Cf. Shapiro [1983*b*]. Where $\psi(r, p)$ abbreviates the synthetic statement that r is a sum of equally spaced, linearly ordered points with initial point p (and hence can serve to model the natural numbers), and where $\text{Con}_{T_{syn}}(r, p)$ abbreviates a synthetic statement of the consistency of T_{syn}, in T_{ext} it cannot be proved that

$$\psi(r, p) \supset [\text{Con}_{T_{syn}}(r, p) \equiv \text{Con}_{T_{syn}}(\omega, 0)] \qquad (*)$$

(where $\text{Con}_{T_{syn}}(\omega, 0)$ abbreviates the ordinary abstract mathematical statement of T_{syn}'s consistency. For, since T_{ext} *can* prove $\text{Con}_{T_{syn}}(\omega, 0)$, if (*) were T_{ext}-provable, $\psi(r, p) \supset \text{Con}_{T_{syn}}(r, p)$ would be also. But by conservativeness, this latter would then be provable in T_{syn} also, contrary to Gödel's (second) incompleteness theorem.

[43] Cf. Malament [1982].

such objects can enter into causal relations, or how they can be epistemically accessible at all. Moreover, as recent studies have brought out, Leibnizian indiscernibility arguments can be constructed, even in the context of space-times of variable curvature, which prove an embarrassment to space-time substantivalism (allowing quantification over space-time points and/or regions). For example, because of invariance of dynamical laws (such as Einstein's field equations) under diffeomorphic transformations of a space-time manifold in itself, there is a strong case that substantivalism is committed to a radical physical indeterminism a priori.[44] Thus, there are serious questions as to whether synthetic alternative theories represent a real advance over standard platonist physics from this broader philosophical perspective.

On the other hand, it may be possible to frame and support synthetic determination claims without space-time substantivalism. A detailed examination of this question cannot be undertaken here. But, especially if "true-of" relations are employed in the modal translates together with (countably) infinitely many special synthetic predicates (as illustrated in the example of § 4, above), the synthetic bases that become available are potentially very rich indeed. Further inquiry along these lines would be needed to settle the matter.

In any case, from the realist perspective outlined above, the whole question appears *recherché*. For real material properties and relations need not be expressible in our languages in the first place. In fact, even the division of "the world" into objects and properties (including relations) can be viewed as in part an imposition of our own thought. Our most neutral descriptions of objective reality are couched in the least informative, general terms, in phrases such as "material reality" or "the actual world". But, despite their relative vacuity, such phrases have their purposes. One of these is to indicate awareness of an objective realm, in no sense created by our thought; and this, in turn, carries an implication that this realm may well not correspond in any direct and simple way with our symbolic efforts to describe it. From this perspective, there is no need to insist on a synthetic vocabulary to pick out a determining set of features of material reality, for it is recognized from the start that such a vocabulary may not be forthcoming. Perhaps, as Bohm and others have envisioned,[45] any conditions we can formulate symbolically will be inadequate due

---

[44] See Earman and Norton [1987].
[45] See Bohm [1957].

to the variety and complexity of our world. If it were adequate—if, for instance, true synthetic determination claims could be formulated in every case—this would have the character of a lucky accident. It should not be required as a way of "making sense of" our applied mathematical modality. And it needn't be required, for it already makes sense to entertain hypothetical mathematical structures which serve merely to represent our mathematical reasoning about the natural world.

Let us finally return to the matter of indispensability arguments. On the modal view, these retain their usual importance, but their import must be reinterpreted. For rather than supporting ordinary (actualist) existence claims, what they support are the modal-existence postulates appealed to in applied mathematics. As we have seen, just what these are depends on the application, and, in principle we draw no line limiting the level of abstractness of the structures whose possibility may receive confirmation through application (coupled with "reverse mathematical" proofs that weaker assumptions are insufficient). Nor have we insisted that such scientific confirmation is the only sort that such assumptions can receive. That is a large question that cannot be resolved here.

However, once the cogency of the modal postulates is granted, the usual platonistic invocation of scientific indispensability (of, for example, a fixed universe of sets) no longer stands firm. For the very possibility of carrying out modal structural mathematics—applied and pure—demonstrates that the ordinary (actualist) existence axioms are not really required. It is sufficient to entertain, for example, the possibility of reifying "results of collecting" along the lines of the msi of set theory, arriving at the possibility of structures characterized in complete abstraction of any "nature" of its component objects. (And, as we have already suggested, if it is replied that mere possibility is all that the "ordinary existence claims of platonist mathematics" come to, then, in fact, a modal interpretation is implicitly being endorsed.) Such possibility claims are logically weaker than the ordinary existence claims associated with objects-platonism; yet they are sufficient for carrying out all the reasoning and constructions of platonist mathematics. Thus, actual platonist mathematical objects are dispensable after all.

As has been recognized, however, the modal-existence claims raise questions of their own. We see no way of explaining them away as linguistic conventions, or of otherwise reducing them to a level of observation, computation, or formal manipulation. At best, the modal approach involves a trade-off *vis-à-vis* standard platonism, and is far from a final resolution of deep philosophical issues in this corner of the foundations of mathematics. Our aim has been not so much to defend as to explore, and if the results are to expose weaknesses as well as strengths, that is in the spirit of our undertaking.

Bibliography

ALLEN, W. [1971], *Getting Even* (New York: Warner).

BARWISE, J. [1977], 'An Introduction to First Order Logic', in J. Barwise, (ed.), *Handbook of Mathematical Logic* (Amsterdam: North Holland), pp. 5–46.

BELINFANTE, F. J. [1973], *A Survey of Hidden-Variables Theories* (Oxford: Pergamon).

BELNAP, N. [unpublished], *Notes on the Science of Logic*.

BENACERRAF, P. [1973], 'Mathematical Truth', in Benacerraf and Putnam [1983], pp. 403–20.

——[1965], 'What Numbers could not be', in Benacerraf and Putnam [1983], pp. 272–94.

——and PUTNAM, H. (eds.) [1983], *Philosophy of Mathematics*, 2nd edn. (Cambridge: Cambridge University Press).

BLUMENTHAL, L. [1961], *A Modern View of Geometry* (New York: Dover).

BOHM, D. [1957], *Causality and Chance in Modern Physics* (Philadelphia, Pa.: University of Pennsylvania Press).

BOOLOS, G. [1971], 'The Iterative Conception of Set', reprinted in Benacerraf and Putnam [1983], pp. 486–502.

——[1985], 'Nominalist Platonism', in *Philosophical Review*, 94: 327–44.

——[1987], 'The Consistency of Frege's *Foundations of Arithmetic*', in J. J. Thomson (ed.), *On Being and Saying* (Cambridge, Mass.: MIT), pp. 3–20.

BUCHHOLZ, W., FEFERMAN, S., POHLERS, W., and SIEG, W. [1981], *Iterated Inductive Definitions and Subsystems of Analysis* (Berlin: Springer).

BURGESS, J. [1983]. 'Why I am not a Nominalist', *Notre Dame Journal of Formal Logic*. 24/1: 93–103.

——[1984], 'Synthetic Mechanics', *Journal of Philosophical Logic*, 13: 379–95.

——[unpublished], 'Sources on the Foundations of Set Theory'.

CARTAN, E. [1924], 'Sur les variétés à connexion affiné et la théorie de la relativité généralisée', *Ann. École Norm. Sup.* 40 (1923), pp. 325–412; and 41 (1924), pp. 1–25.

CHIHARA, C. S. [1973], *Ontology and the Vicious Circle Principle* (Ithaca, NY.: Cornell University Press).

CHUAQUI, R. [1972], 'Forcing and the Impredicative Theory of Classes', *Journal of Symbolic Logic*, 37, 37/1: 1–18.

CLAUSER, J., and SHIMONY, A. [1978], 'Bell's Theorem: Experimental Tests and Implications', *Reports on Progress in Physics*, 41: 1882–927.

COCCHIARELLA, N. B. [1966], 'A Logic of Actual and Possible Objects', *Journal of Symbolic Logic*, 31: 688 ff.

—— [forthcoming], 'Philosophical Perspectives on Formal Theories of Predication', in D. Gabbay and F. Guenthner (eds.), *Handbook of Philosophical Logic*, vol. iv (Dordrecht, The Netherlands: Reidel).

—— [1975], 'On the Primary and Secondary Semantics of Logical Necessity', *Journal of Philosophical Logic*, 4: 13–27.

—— [1984], 'Philosophical Perspectives on Quantification in Tense and Modal Logic', in D. Gabbay and F. Guenthner (eds.), *Handbook of Philosophical Logic*, vol. ii (Dordrecht, The Netherlands: Reidel), pp. 309–53.

—— [1986], *Logical Investigations of Predication Theory and the Problem of Universals* (Naples: Bibliopolis).

COHEN, P. J. [1966], *Set Theory and the Continuum Hypothesis* (New York: Benjamin).

CORCORAN, J. [1980], 'Categoricity', *History and Philosophy of Logic*, 1: 187–207.

CRAIG, W. [1953], 'On Axiomatizability within a System', *Journal of Symbolic Logic*, 18/1: 30–2.

DEDEKIND, R. [1901], 'The Nature and Meaning of Numbers', in *Essays on the Theory of Numbers* (New York: Dover, 1963), 31–115, transl. from the German original, "Was sind und was sollen die Zahlen" (Brunswick: Vieweg, 1888).

DIEUDONNÉ, J. [1969], *Foundations of Modern Analysis* (New York: Academic Press).

DRAKE, F. R. [1974], *Set Theory: An Introduction to Large Cardinals* (Amsterdam: North Holland).

DUMMETT, M. [1963], 'The Philosophical Significance of Gödel's Theorem', in M. Dummett, *Truth and Other Enigmas* (Cambridge, Mass.: Harvard University Press, 1978), pp. 186–201.

—— [1977], *Elements of Intuitionism* (Oxford: Oxford University Press).

—— [1978], 'Platonism', in M. Dummet, *Truth and Other Enigmas* (Cambridge, Mass.: Harvard University Press, 1978), pp. 202–14.

EARMAN, J., and NORTON, J. [1987], 'What Price Substantivalism: The Hole Story', *British Journal for the Philosophy of Science*, 38: 515–25.

FEFERMAN, S. [1978], 'The Logic of Mathematical Discovery vs. the Logical Structure of Mathematics', in P. D. Asquith and I. Hacking, *PSA 1978*, vol. ii (East Lansing, Mich.: Philosophy of Science Association, 1981), pp. 309–327.

—— [1988], 'Hilbert's Programme Relativized: Proof Theoretical and Foundational Reductions', *Journal of Symbolic Logic*, 53: 164–84.

FIELD, H. [1980], *Science without Numbers* (Princeton, NJ: Princeton University Press).

—— [1984], 'Is Mathematical Knowledge just Logical Knowledge', *Philosophical Review*, 93: 509–52.

—— [1985], 'On Conservativeness and Incompleteness', *Journal of Philosophy*, 82: 239–60.

FINE, K. [1981], 'First-order Modal Theories, I—Sets', *Noûs*, 15/2: 177–205.

FRAENKEL, A., BAR-HILLEL, Y., and LÉVY, A. [1973], *Foundations of Set Theory*, 2nd edn. (Amsterdam: North Holland).

FREGE, G. [1978], *The Foundations of Arithmetic* (Oxford: Basil Blackwell).

FRIEDMAN, H. [1981], 'On the Necessary Uses of Abstract Set Theory', *Advances in Mathematics*, 41: 209–80.

FRIEDMAN, M. [1983], *Foundations of Space-Time Theories* (Princeton, NJ: Princeton University Press).

GALLEN, D. [1975], *Intensional and Higher-Order Modal Logic* (Amsterdam: North Holland).

GLEASON, A. [1957], 'Measures on the Closed Subspaces of a Hilbert Space', *Journal of Mathematics and Mechanics*, 6: 885–93.

GÖDEL, K. [1947], 'What is Cantor's Continuum Problem?', in Benacerraf and Putnam [1983], pp. 470–85.

GOODMAN, N. [1955], *Fact, Fiction, and Forecast* (Cambridge, Mass.: Harvard University Press).

—— [1972], 'The Way the World Is', in *Problems and Projects* (Indianapolis, Ind.: Bobbs-Merrill) pp. 24–32.

—— [1977], *The Structure of Appearance*, 3rd edn. (Dordrecht, The Netherlands: Reidel).

—— [1978], *Ways of Worldmaking* (Indianapolis, Ind.: Hackett).

HALE, S. C., and RESNIK, M. D. [1987], 'Science Nominalized', *Philosophy of Science*, 54: 277–80.

HALLETT, M. [1984], *Cantorian Set Theory and Limitation of Size* (Oxford: Oxford University Press).

HARRINGTON, L. A., MORLEY, M. D., SCEDROV, A., and SIMPSON, S. G. [1985], *Harvey Friedman's Research on the Foundations of Mathematics* (Amsterdam: North Holland).

HELLMAN, G. [1983], 'Realist Principles', *Philosophy of Science*, 50: 227–49.

—— [1985], 'Determination and Logical Truth', *Journal of Philosophy*, 82: 607–16.

—— and THOMPSON, F. W. [1975], 'Physicalism: Ontology, Determination, and Reduction', *Journal of Philosophy*, 72/17: 551–64.

HICKS, N. J. [1971], *Notes on Differential Geometry* (New York: Van Nostrand).

HILBERT, D. [1971], *Foundations of Geometry* (La Salle, Ill.: Open Court).

HISCUS, A. [unpublished], 'Relativity Principles' (Indiana University, Ph.D. thesis).

HODES, H. [1984], 'Axioms for Actuality', *Journal of Philosophical Logic*, 13: 27–34.

HORGAN, T. [1984], 'Science Nominalized', *Philosophy of Science*, 51: 529–49.

HORGAN, T. [1987], 'Science Nominalized Properly', *Philosophy of Science*, 54: 281–2.

JAUCH, J. M. [1968], *Foundations of Quantum Mechanics* (Reading, Mass.: Addison-Wesley).

JECH, T. [1978], *Set Theory* (New York: Academic Press).

KANAMORI, A., and MAGIDOR, M. [1978], 'The Evolution of Large Cardinal Axioms in Set Theory', in G. H. Muller and D. S. Scott (eds.), *Higher Set Theory*, *Springer Lecture Notes in Mathematics*, vol. 669 (Berlin: Springer).

KESSLER, G. [1978], 'Mathematics and Modality', *Noûs*, 12: 421–41.

KIM, J. [1984], 'Concepts of Supervenience', *Philosophy and Phenomenological Research*, 45: 153–76.

KITCHER, P. [1983], *The Nature of Mathematical Knowledge* (Oxford: Oxford University Press).

KRANTZ, D., LUCE, R., SUPPES, P., and TVERSKY, A. [1971], *Foundations of Measurement*, vol. i (New York: Academic Press).

KREISEL, G. [1967], 'Informal Rigour and Completeness Proofs', in I. Lakatos (ed.), *Problems in the Philosophy of Mathematics* (Amsterdam: North Holland), pp. 138–86.

—— [1972], Review of Putnam [1967], *Journal of Symbolic Logic*, 37: 402–4.

KRIPKE, S. [1963], 'Semantical Considerations on Modal Logics', *Acta Philosophica Fennica: Modal and Many-Valued Logics*, pp. 83–94.

LÉVY, A. [1960], 'Axiom Schemata of Strong Infinity in Axiomatic Set Theory', *Pacific Journal of Mathematics*, 10: 223–38.

LEWIS, D. K. [1973], *Counterfactuals* (Oxford: Oxford University Press).

MCCARTHY, T. [1986], 'Platonism and Possibility', *Journal of Philosophy*, 83: 275–90.

MADDY, P. [1980], 'Perception and Mathematical Intuition', *Philosophical Review*, 89: 163–96.

—— [1983], 'Proper Classes', *Journal of Symbolic Logic*, 48: 113–39.

—— [1988], 'Believing the Axioms', Part 1, *Journal of Symbolic Logic*, 53: 481–511.

MALAMENT, D. [1982], 'Hartry Field's *Science Without Numbers*', *Journal of Philosophy*, 79: 523–34.

—— [unpublished], Lecture Notes on General Relativity.

MENZEL, C. [1986], 'On the Iterative Explanation of the Paradoxes', *Philosophical Studies*, 49: 37–61.

MISNER, C. W., THORNE, K. S., and WHEELER, J. A. [1973], *Gravitation* (San Francisco, Calif.: Freeman).

MONTAGUE, R. [1965], 'Set Theory and Higher Order Logic', in J. Crossley and M. Dummett (eds.), *Formal Systems and Recursive Functions* (Amsterdam: North Holland), pp. 131–48.

MOORE, G. H. [1980], 'Beyond First-Order Logic: The Historical Interplay between Mathematical Logic and Axiomatic Set Theory', *History and Philosophy of Logic*, 1: 95–137.

OHANIAN, H. C. [1976], *Gravitation and Space-Time* (New York: Norton).

PARIS, J., and HARRINGTON, L. [1977], 'A Mathematical Incompleteness in Peano Arithmetic', in Barwise [1977], pp. 1133–42.

PARSONS, C. [1983], *Mathematics in Philosophy* (Ithaca, NY.: Cornell University Press).

——[forthcoming], 'The Structuralist View of Mathematical Objects', *Synthese*.

PIRON, C. [1976], *Foundations of Quantum Physics* (New York: Benjamin).

POST, J. [1987], *The Faces of Existence* (Ithaca, NY: Cornell University Press).

PRUGOVECKI, E. [1981], *Quantum Mechanics in Hilbert Space* (New York: Academic Press).

PUTNAM, H. [1965], 'Craig's Theorem', in *Philosophical Papers*, 1 (1975): 228–36 (Cambridge: Cambridge University Press).

——[1967], 'Mathematics without Foundations', in Benacerraf and Putnam [1983], pp. 295–311.

——[1970], 'On Properties', in *Philosophical Papers*, 1 (1975): 305–22.

——[1975], *Philosophical Papers*, 1 (Cambridge: Cambridge University Press).

——[1981], *Reason, Truth, and History* (Cambridge: Cambridge University Press).

QUINE, W. V. [1961], *From a Logical Point of View* (New York: Harper).

——[1969], *Ontological Relativity and Other Essays* (New York: Columbia University Press).

——[1976], *The Ways of Paradox and Other Essays* (Cambridge, Mass.: Harvard University Press).

REINHARDT, W. [1974*a*], 'Remarks on Reflection Principles, Large Cardinals, and Elementary Embedding', in T. Jech (ed.), *Axiomatic Set Theory* (Proceedings of Symposia in Pure Mathematics, Vol. 13, Part 2), pp. 189–205 (Providence, RI: American Mathematical Society).

——[1974*b*], 'Set Existence Principles of Shoenfield, Ackermann, and Powell', *Fundamental Mathematicae*, 84: 5–34.

RESNIK, M. [1981], 'Mathematics as a Science of Patterns: Ontology and Reference', *Noûs*, 15: 529–50.

——[1985], 'Ontology and Logic: Remarks on Hartry Field's Anti-Platonist Philosophy of Mathematics', *History and Philosophy of Logic*, 6: 191–209.

ROBBIN, J. [1969], *Mathematical Logic* (New York: Benjamin).

RUSSELL, B. A. W. [1919], *Introduction to Mathematical Philosophy* (London: Allen & Unwin).

SCOTT, D. [1974], 'Axiomatizing Set Theory', in T. Jech (ed.), *Axiomatic Set Theory* (Proceedings of Symposia in Pure Mathematics, Vol. 13, Part 2), pp. 207–14 (Providence, RI: American Mathematical Society).

SHAPIRO, S. [1983*a*], 'Mathematics and Reality', *Philosophy of Science*, 50: 523–48.

—— [1983*b*], 'Conservativeness and Incompleteness', *Journal of Philosophy*, 80: 521–31.

—— [1985], 'Second Order Languages and Mathematical Practice', *Journal of Symbolic Logic*, 50: 714–42.

SHEPERDSON, J. C. [1951–3], 'Inner Models for Set Theory', *Journal of Symbolic Logic*, 16 (1951): 161–90; 17 (1952): 225–37; 18 (1953): 145–67.

SHOENFIELD, J. [1967], *Mathematical Logic* (Reading, Mass.: Addison-Wesley).

—— [1977], 'Axioms of Set Theory', in Barwise [1977], pp. 321–44.

STALNAKER, R. C. [1968], 'A Theory of Conditionals', in W. L. Harper, R. Stalnaker, and G. Pearce (eds.), *Ifs* (Dordrecht, The Netherlands: Reidel), pp. 41–55.

STEIN, H. [1988], '*Logos*, Logic and *Logistike*: Some Philosophical Remarks on 19th Century Transformation of Mathematics', in W. Aspray and P. Kitcher (eds.), *Minnesota Studies in Philosophy of Science*, vol. xi: *History and Philosophy of Modern Mathematics* (Minneapolis, Minn.: University of Minnesota Press).

TAIT, W. W. [1983], 'Against Intuitionism: Constructive Mathematics is a Part of Classical Mathematics', *Journal of Philosophy*, 12: 173–95.

—— [1986*a*], 'Truth and Proof: The Platonism of Mathematics', *Synthese*, 69: 341–70.

—— [1986*b*]. 'Critical Notice: Charles Parsons' *Mathematics in Philosophy*', *Philosophy of Science*, 53: 588–607.

—— [forthcoming], 'The Iterative Conception of Set'.

TELLER, P. [1983], 'A Poor Man's Guide to Supervenience and Determination', *Southern Journal of Philosophy*, 12, Supplement: 137–62.

TRAUTMAN, A. [1966], 'Comparison of Newtonian and Relativistic Theories of Space-Time', in B. Hoffmann (ed.), *Perspectives in Geometry and Relativity* (Bloomington, Ind.: Indiana University Press).

WESTON, T. [1976], 'Kreisel, the Continuum Hypothesis and Second-Order Set Theory', *Journal of Philosophical Logic*, 5: 281–98.

ZERMELO, E. [1930], 'Über Grenzzahlen und Mengenbereiche: neue Untersuchungen über die Grundlagen der Mengenlehre', *Fundamenta Mathematicae*, 16: 29–47.

Index